# THE BIOLOGY OF BANANAS

First American Edition 2020
Kane Miller, A Division of EDC Publishing



For information contact:
Kane Miller, A Division of EDC Publishing
PO Box 470663
Tulsa, OK 74147-0663
www.kanemiller.com
www.edcpub.com
www.usbornebooksandmore.com

Library of Congress Control Number: 2019936215
Printed in China
ISBN: 978-1-68464-004-1

1 2 3 4 5 6 7 8 9 10

**SAFETY NOTICE**: The experiments in this book must be conducted under adult supervision, and with all reasonable caution including an awareness of food allergies and intolerance. The instructions provided in each experiment are no replacement for the sound judgment of participants. The author and publisher accept no liability for any mishap or injury arising from participation in the experiments.

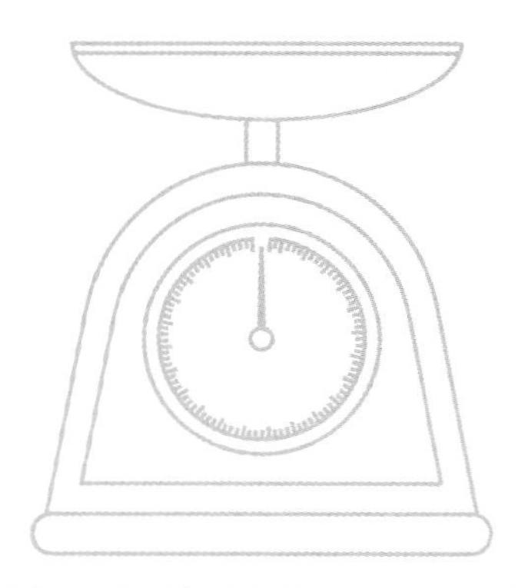

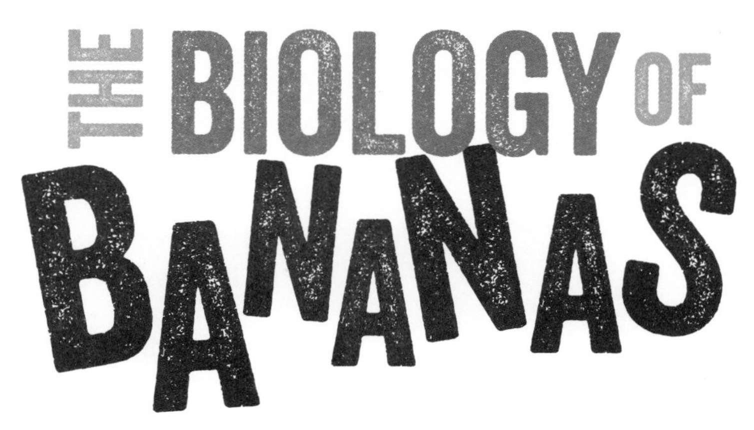


## THE CURIOUS WORLD OF **KITCHEN SCIENCE**

KATIE STECKLES

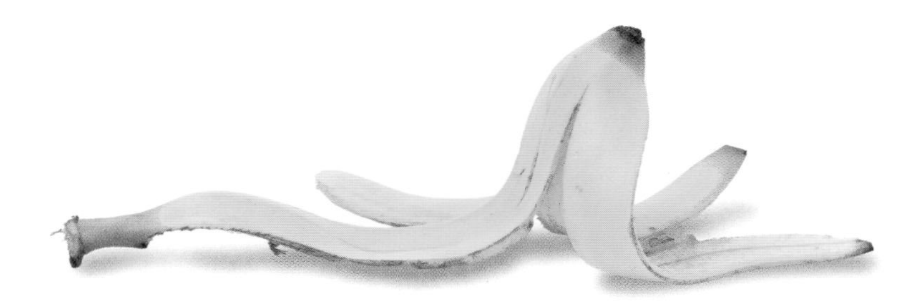

**Kane Miller**
A DIVISION OF EDC PUBLISHING

# CONTENTS

## CHAPTER 3: YOU



## CHAPTER 4: THE WORLD

# INTRODUCTION

Biology is the study of living things, like plants and animals (including people!). It covers everything from tiny cells to huge trees and animals, all the way up to global ecosystems. Being a biologist involves knowing a little chemistry, a little math, and a little physics, and combining these things to study complicated systems that interact in fascinating ways.

You will find lots of examples of biology in your own kitchen—including the food you're cooking and eating, and how it's processed by your body. This book will show you some of the ways you can study biology without leaving the kitchen, through interesting biological facts, questions to test your learning, and experiments for you to try at home.

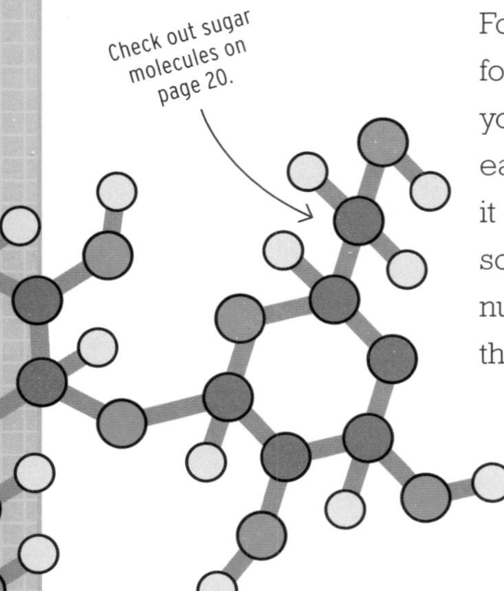


This book has four chapters: Plants, Food, You, and The World. Most of the food you eat comes from plants—even if you are eating meat, the animal you're eating will have been fed plants to make it grow, and the Plants section looks at some of the ways plants provide nutritious and delicious food, and what they need to grow.

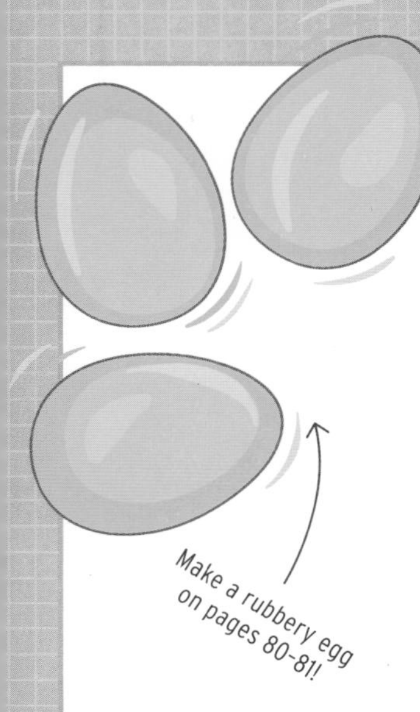


The Food chapter looks at the biology of food itself, biological processes that go on inside your food, and some of the foods you can make using biological processes.

Much of the biology that goes on in your kitchen takes place inside your body. The You chapter looks at how your body processes and uses the energy from food, and some of the things you need to think about when deciding what to cook for dinner.

Biology also extends beyond your kitchen, and beyond the garden outside—the processes that create the plants and food we eat are part of vast ecosystems that connect across the planet. The World looks at the effect the food you eat has on the environment, and how scientists are helping to reduce the impact of the important business of feeding the world's population.

Each chapter also includes experiments and quizzes. Some quiz questions will cover information you just read, and some will not—the answers at the back of the book explain and offer more information on both. Remember, there is much more information on these topics available, so if you find something interesting, go and find out more about it!

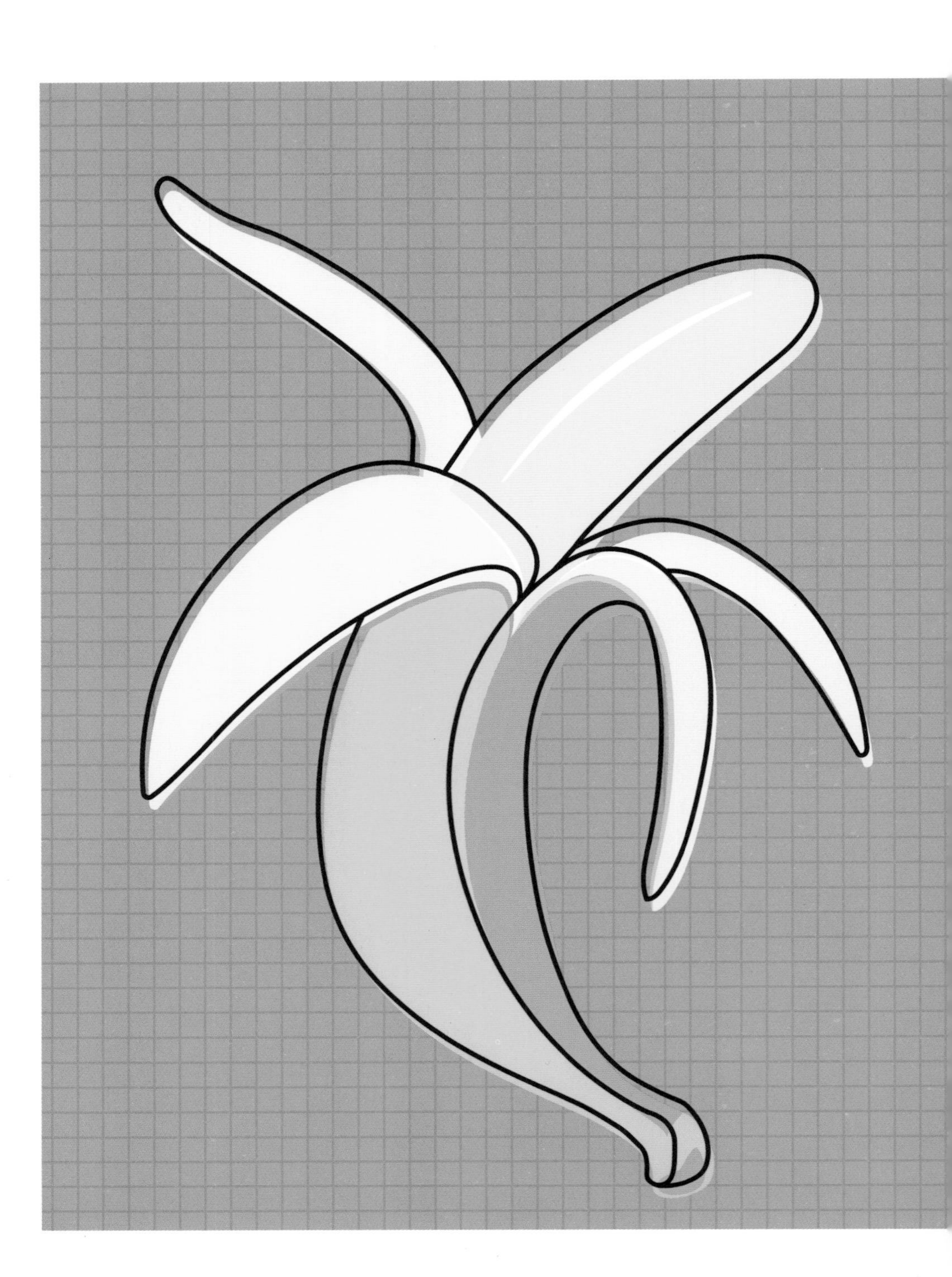

# CHAPTER 1
# PLANTS

DISCOVER...

LEARN...

EXPERIMENT...

# DISCOVER: THE BIOLOGY OF A BANANA

**You might eat a banana without thinking about where it comes from, but, biologically, bananas are very interesting. Here are some banana facts, which you might find surprising.**

- **Bananas don't grow on trees**
The banana plant doesn't have a wooden trunk like a tree; it's made of banana leaves, which grow and curl around each other to form a stem.

- **Bananas are technically berries**
Botanically, a berry has an outer skin called the exocarp, a fleshy middle inside (the mesocarp), and then multiple seeds in the endocarp (the part of the flesh in the very center). Blueberries, kiwifruits, and even eggplants are berries. A cherry, with one large stone, isn't a berry, and neither is a strawberry (because its seeds are on the outside) or a raspberry (which has many small sections, called drupes, each containing a single seed).

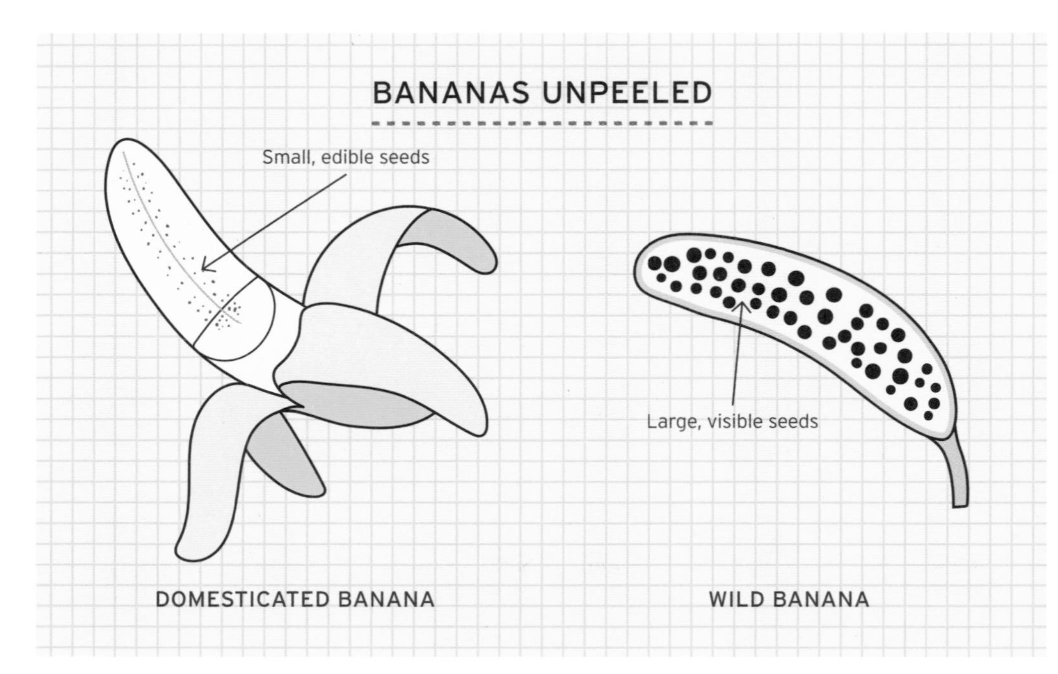

• **Banana plants don't grow naturally** Wild bananas are very different than domesticated ones. In particular, they have huge inedible seeds, and banana growers created more edible varieties by always choosing the plants with smaller seeds—until they became the tiny black spots you see inside bananas today. This means bananas can't be grown by planting seeds in the ground; they're grown from offshoots, and now almost all bananas are descended from two original species, *Musa acuminata* and *Musa balbisiana*. They're both very vulnerable to disease.

### BANANA-NOT

Banana-flavored candies often don't taste like the bananas we eat at home. That's because they're not made with actual bananas, but instead use chemicals called esters, which mimic the flavor. Some varieties of bananas, including the rare Gros Michel, have a sweeter taste that is closer to the flavor you get in banana candies. (Find out more about esters on pages 118-119!)

• **Bananas grow pointing upward!** While you imagine bananas hanging downward on a plant, the stem of the banana is actually at the bottom, and they grow upward in bunches that can weigh over 45 kg (100 lb.).

## CAUTION: BANANAS ARE RADIOACTIVE!

Bananas contain high levels of potassium, which is slightly radioactive. There's not enough radioactivity in a banana to hurt you, but you can measure other radioactive sources in terms of the Banana Equivalent Dose: the number of bananas you'd have to eat to get the same dose of radiation.

- You're exposed to about 100 bananas of radiation per day.
- A CT scan is about 70,000 bananas.
- The radiation in 3,500,000 bananas is enough to kill you.

# **DISCOVER:** THE STRUCTURE OF CELLS

**In biology, cells are the building blocks of everything: plants, animals, fungi (like yeast), and bacteria are all made from different types of cells. Here are the main kinds of cell you might see.**

## PARTS OF THE CELL

- **Nucleus:** This is found in all plant and animal cells. It's where the cell's DNA is stored, and it functions as the "brain" of the cell, giving it instructions about what to do and how to grow.
- **Cytoplasm:** All cells are filled with it—a jellylike fluid, made mostly of water and salt, which fills the cell and holds all the other parts in place.
- **Cell membrane:** All cells have one. It wraps around the outside and holds everything in. It's made from a double-thick layer of proteins and fat molecules.
- **Mitochondria:** These are tiny organelles—specialized structures suspended in the cytoplasm. Mitochondria produce energy for the cell by converting oxygen into carbon dioxide, known as respiration.
- **Cell wall:** Plant cells are surrounded by a thick wall made from cellulose—a type of sugar—which holds the shape of the cell rigid. It's the reason why plant stems are stiff and can hold themselves upright. Some bacteria

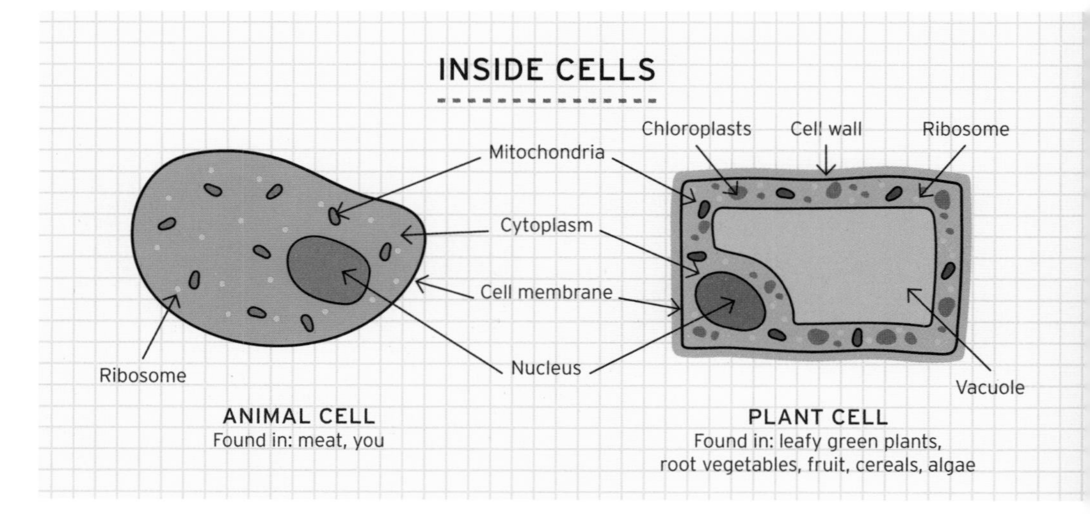

and yeast cells have a different type of cell wall, made from a mixture of proteins and sugars.

• **Vacuole:** This is a bubble inside a plant cell used for storing nutrients or waste products. When a plant hasn't gotten enough water, less fluid is stored in the vacuoles, so the cells get smaller and the plant droops and wilts.

• **Chloroplasts:** Only found in plant cells, these contain molecules of a green substance called chlorophyll, which is why plants look green. They're used by the plant cell for photosynthesis—converting energy from light, along with water and carbon dioxide, into sugars the cell needs, and oxygen that's released into the air.

• **Ribosome:** These are tiny particles of protein used to join together molecules called amino acids to make other proteins. They follow instructions from the nucleus to work out what to make.

• **Plasmids:** Found in bacteria cells, these are circular loops of DNA. Since the bacteria cell doesn't contain a nucleus, its DNA is stored loose in the cytoplasm.

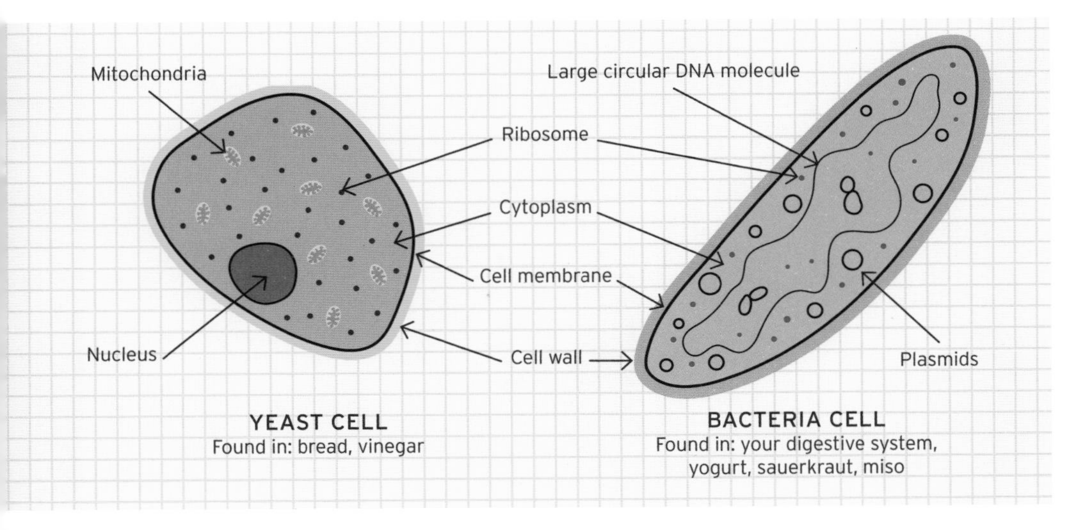

# LEARN ABOUT: BANANAS

Even if you eat bananas every day, you might find there are things you don't know about them. For instance, a lot of people are surprised to hear that a banana is technically a berry. (See page 29 for more about the different categories fruits fall into.) See how much you know, or can guess, about bananas in this multiple-choice quiz.

## POP QUIZ: BANANAS

**1.** What happens when you put a banana in water?
a) It will float
b) It will sink
c) It will dissolve

**2.** How big are wild bananas compared to the domesticated varieties we eat?
a) Bigger
b) Smaller
c) The same size

**3.** Which of these metals is found in a banana?
a) Iron
b) Potassium
c) Magnesium
d) All of the above

**4.** What is a banana plant's trunk made from?
a) Wood
b) Bamboo
c) Banana leaves

**5.** Which of these is a term used to refer to a bunch of 10-20 bananas?
a) Hand
b) Foot
c) Squid

**6.** What proportion of our DNA do humans share with bananas?
a) 10%
b) 50%
c) 90%

**7.** Which of the following is it claimed banana peel can do?
a) Relieve itching
b) Whiten teeth
c) Attract butterflies
d) Polish shoes
e) Remove splinters
f) All of the above

**8.** Which of these is not part of a banana?
a) Endocarp
b) River carp
c) Exocarp

**9.** What's the world record for the most bananas peeled and eaten in one minute?
a) 8
b) 17
c) 25

**10.** Which of these fruits is in the same botanical category of fruit as a banana?
a) Cherry
b) Blueberry
c) Raspberry

# **LEARN ABOUT:** PARTS OF A CELL

**Which of the following are found in which types of cells? Look at the diagrams on pages 12–13 if you're not sure, and check the boxes.**

## CHECK THE BOXES

| CELL PART | ANIMAL CELL | PLANT CELL | YEAST CELL | BACTERIA CELL |
|---|---|---|---|---|
| MITOCHONDRIA | | | | |
| CHLOROPLASTS | | | | |
| CELL MEMBRANE | | | | |
| NUCLEUS | | | | |
| CELL WALL | | | | |
| PLASMIDS | | | | |
| CYTOPLASM | | | | |
| RIBOSOMES | | | | |
| VACUOLE | | | | |

Which types of cell (animal, plant, yeast, or bacteria) would you expect to find in each of these foods, or use in making them? There might be more than one. **1.** Roasted ham **2.** Dressed salad **3.** Cheeseburger with lettuce **4.** Strawberry yogurt **5.** Chicken wings

# EXPERIMENT: MAKING CELL COOKIES

A great (and delicious) way to remember all the parts of a cell is to decorate some cookies to look like different kinds of cells. Pick your favorite type of cell from pages 12–13—or do them all! Use the diagrams to see what you'll need to include.

## YOU WILL NEED:

- Plain cookies (round, rectangular, or uneven shaped)
- Different colors of ready-made frosting in squeezable tubes
- Different sizes of candy, sprinkles, chocolate chips, and other tasty things to stick on

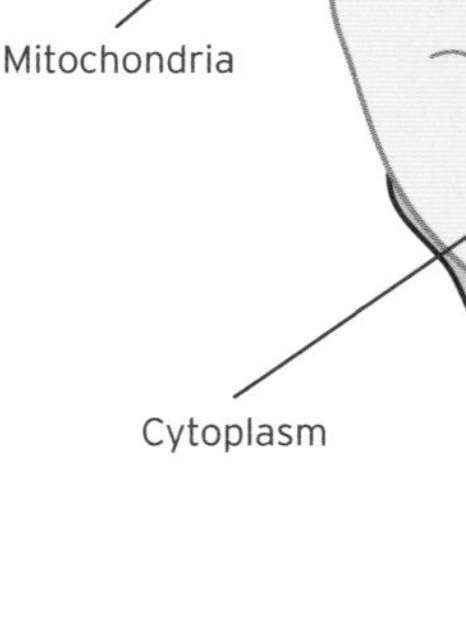


**REMEMBER:** Every cell is unique, so if you're making a whole batch, your cookies don't have to all look the same!

**WHAT TO DO:**

**1.** Decide what type of cell you're going to make—animal, plant, yeast, or bacteria—or maybe make one of each!

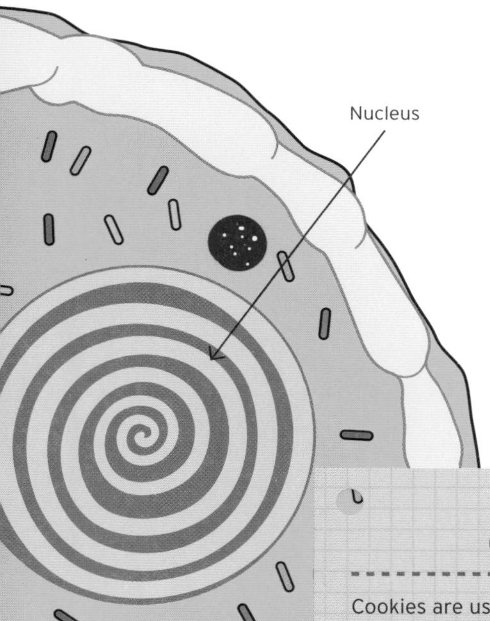


**2.** For each of the parts of the cell you want to make, decide which type of decoration is going to represent it. You could use frosting to pipe a cell wall around the outside, a large candy for the nucleus, green gumdrops for the chloroplasts, a marshmallow for the vacuole, and sprinkles for the tiny ribosomes.

**3.** Use frosting to stick the candies on the cookies in the right arrangement.

**4.** Share and enjoy delicious cell cookies! Ask people if they can identify what all the parts are.

## CELLS IN YOUR CELLS

Cookies are usually made with flour, sugar, butter, and eggs. Most of these ingredients don't include any cells—sugar is made up of sugar molecules, and butter is a mixture of fats and proteins. Flour is made by grinding up wheat kernels, and is a mixture of starches, sugars, and proteins. But the egg contains the most interesting biology (read all about it on pages 78-79).

Along with egg white (which is mainly water, with a little protein), and egg yolk (high in fats and protein), the egg also contains a blastodisc—a small white dot on the surface of the yolk, which contains around 20,000 cells. These are what might have grown into a chicken if the egg had been fertilized and incubated, instead of being made into cookies. So if they were made with egg, your cell cookies could contain the remnants of actual cells!

# EXPERIMENT: LOOKING THROUGH A MICROSCOPE

**Much of what we study in biology is too small to see with the naked eye. The microscope, invented around 1590, allows scientists to see a whole new world of tiny things, including cells and even molecules. But how big are these small things really?**

## MEASURING LENGTHS!

**PINHEAD**
The head of a standard sewing pin is around 1 millimeter across.

**HAIR**
Human hair varies in thickness but is on average about 100 micrometers thick. That means you could fit ten side by side and it would be 1 millimeter across.

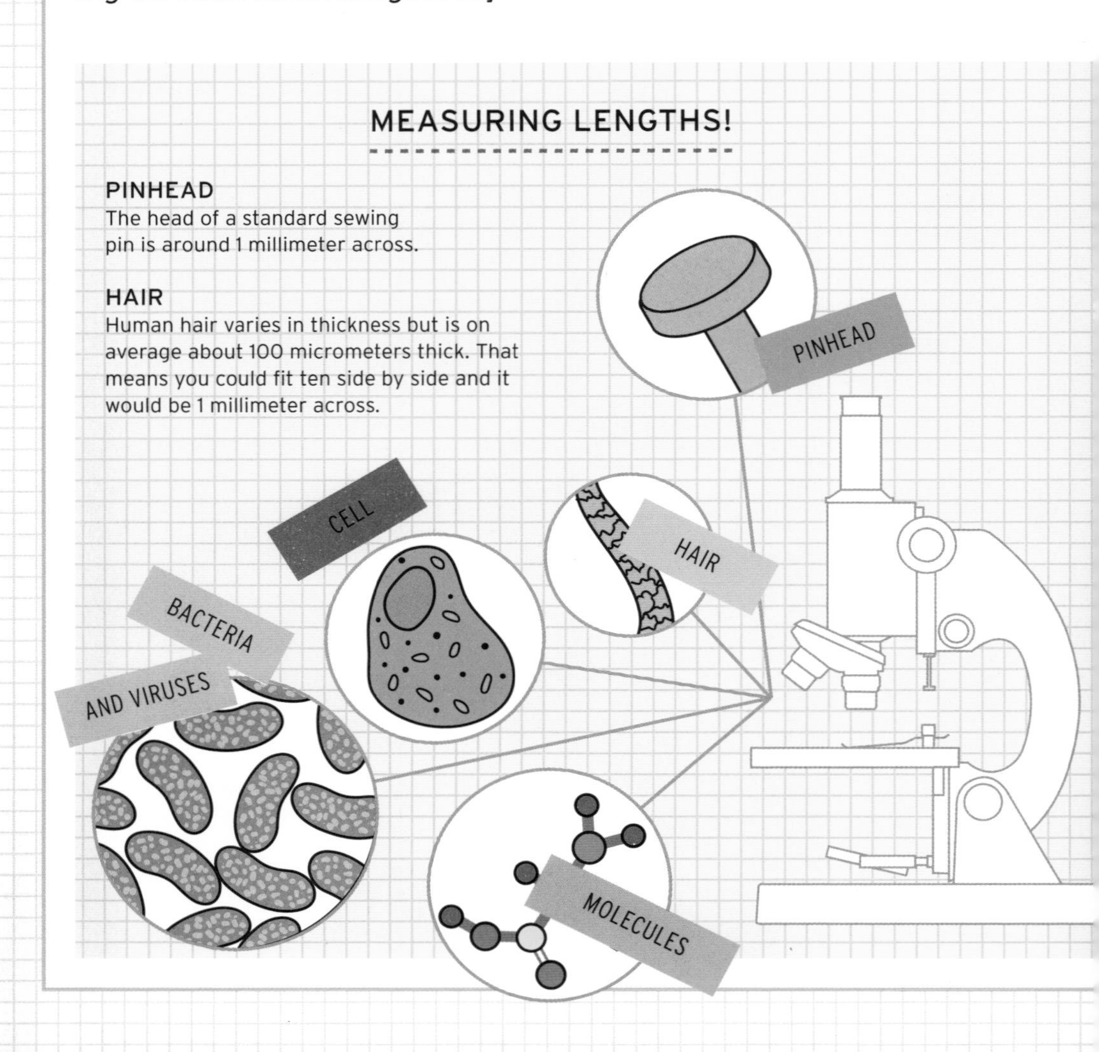

## MAGNIFICATION

Microscopes make things appear bigger, a process called magnification. To magnify something that's 1 micrometer across so that it looks 1 centimeter across, you'll need to magnify it by 10,000 times because there are 10,000 micrometers in a centimeter.

If you have a microscope, or a magnifying glass, you might be able to look at some of the things that make up the foods you eat. Many microscopes can have up to 1000 times magnification—but even if you only have 20 to 40 times, you can still see some interesting things.

### CELL

The largest human body cells–eggs, found in a woman's uterus–are around 100 micrometers across, but most are much smaller. Skin cells are about 30 micrometers in diameter, and a red blood cell, which carries oxygen around in your bloodstream, is around 10 micrometers. A typical yeast cell measures 3-4 micrometers.

### BACTERIA AND VIRUSES

The largest bacteria can measure up to 500 micrometers (half a millimeter), but an average bacterium–such as *E. coli*, which can cause food poisoning if it gets in the wrong place–is around 2 micrometers across. Viruses are much smaller, ranging from 100 nanometers (1/10 of a micrometer) down to 20 nanometers.

### MOLECULES

The DNA molecules inside the cells of your body and food measure around 2 nanometers across, and the amino acids which make up proteins are around 1 nanometer. A single hydrogen atom measures 0.1 nanometers.

### DISCOVER MORE

**YOU COULD LOOK AT:**

- Sugar
- Salt
- Flour
- Seaweed
- Mushroom spores
- Plant leaves

To investigate what parts of you look like close up, you could look at:

- Skin cells from the inside of your cheek
- A strand of hair
- Plaque from your teeth–scrape off some of the white substance that builds up on them

- 1 meter is 100 centimeters (39 1/3 in.)
- 1 centimeter is 10 millimeters (2/5 in.)
- 1 millimeter is 1000 micrometers (1/25 in.)
- 1 micrometer is 1000 nanometers (1/25000 in.)

# **DISCOVER:** WHAT IS CELLULOSE?

**Cellulose, found in plant cell walls, is made up of long chains of sugar molecules called polysaccharides. Like many organic substances, it only contains atoms of carbon, hydrogen, and oxygen, but it's the way the atoms are joined together that gives it its interesting properties.**

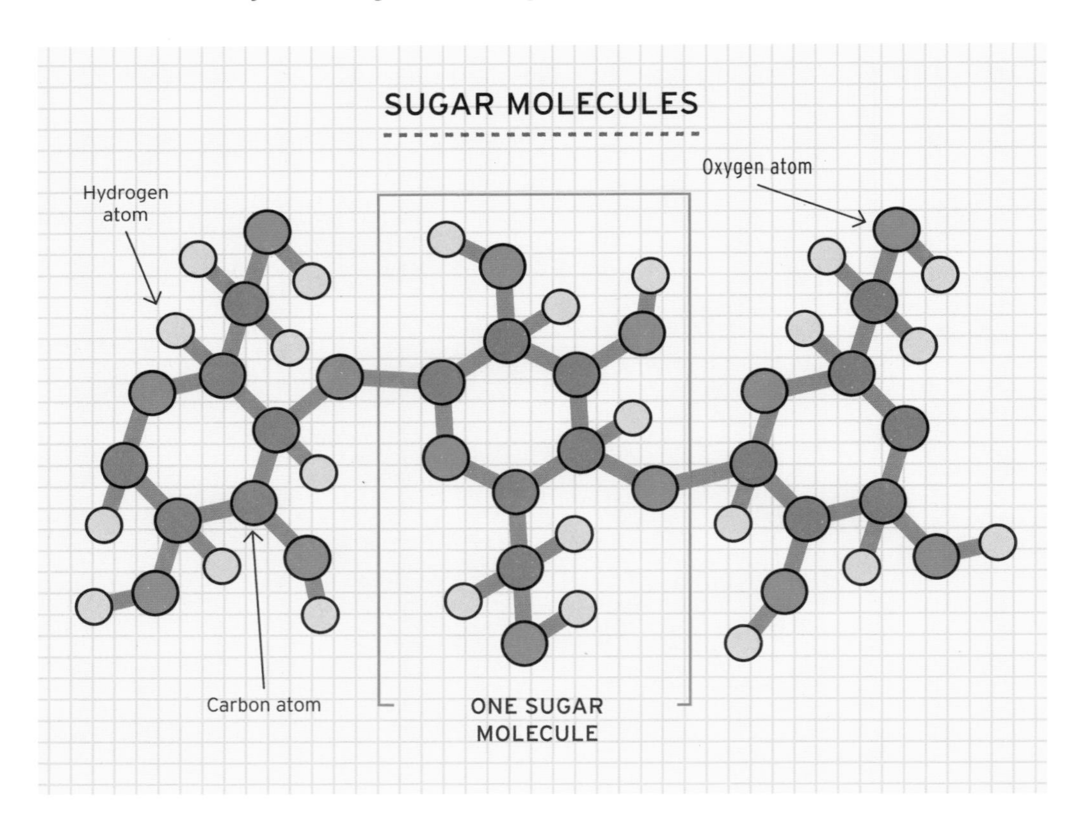


These long chains of molecules make cellulose a strong, fibrous substance. It's what gives the cell walls of plant cells their strength, and it makes up over 90% of cotton fiber. Inside their cells, plants join together sugar molecules to synthesize cellulose, and it's the most abundant organic polymer (long-chain molecule) on the planet.

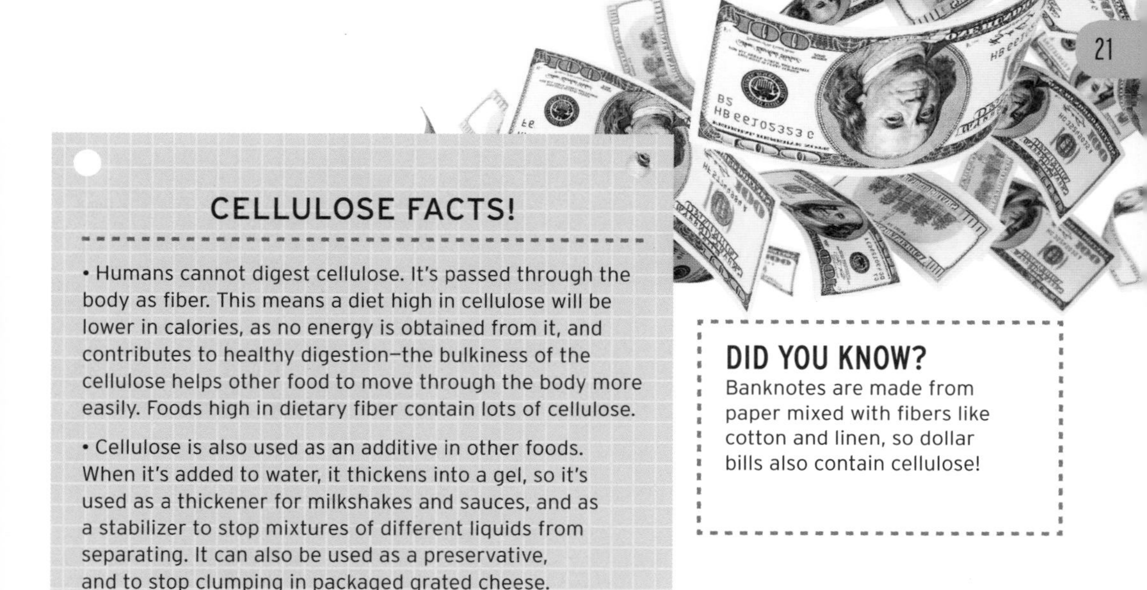

## CELLULOSE FACTS!

• Humans cannot digest cellulose. It's passed through the body as fiber. This means a diet high in cellulose will be lower in calories, as no energy is obtained from it, and contributes to healthy digestion—the bulkiness of the cellulose helps other food to move through the body more easily. Foods high in dietary fiber contain lots of cellulose.

• Cellulose is also used as an additive in other foods. When it's added to water, it thickens into a gel, so it's used as a thickener for milkshakes and sauces, and as a stabilizer to stop mixtures of different liquids from separating. It can also be used as a preservative, and to stop clumping in packaged grated cheese.

• Termites, and ruminants—grazing animals like cows, which have a special digestive system with several stomachs—can digest cellulose. Their stomachs contain species of single-celled organisms that help them break down the molecules. This is why cows and sheep are able to survive by eating grass, which is mostly cellulose. Humans couldn't survive on grass, as we'd get very little nutrition from it as we digested it.

• As well as being part of many foods, and a large constituent of cotton fabric, cellulose is used to make paper and cardboard. Trees are made from plant cells, and wood is 40–50% cellulose—it's made into paper by pulping. It's also used to make cellophane (a kind of plastic used in many types of food packaging) and rayon, an artificial fiber used to make fabric.

• Fungi can also break down cellulose, which is how they're able to rot wood.

### DID YOU KNOW?

Banknotes are made from paper mixed with fibers like cotton and linen, so dollar bills also contain cellulose!

Test your cow knowledge on page 136!

# LEARN ABOUT: HOW MANY CELLS?

**You've learned about the different kinds of cells there are, and you may even have had a chance to see one under a microscope. Plants and animals, and many of the foods you eat, are made from cells—but how many cells might you find in one item of food?**

Cells are incredibly small (and it'd be pretty hard to count them all!), but we can get a rough estimate by weighing things and using a bit of multiplication…

An average cell weighs one nanogram. This means 1,000,000,000 (a billion) cells weigh one gram. So how many cells does this mean we could expect to find in a piece of food? To find out, you will need a kitchen scale that measures in grams and a calculator.

Use your scale to weigh food items that are made from cells—such as a lettuce leaf, a slice of tomato, or a burger patty. (If it's not made of pure meat, it might contain fewer cells. If you can find what percentage of the burger is meat, you could work out a more accurate value.)

## A FEW EXAMPLES

**LETTUCE LEAF:** A leaf might weigh between 5 g (1/5 oz.) and 25 g (1 oz.), so a small leaf could contain 5,000,000,000 (five billion) plant cells, and a large one would have 25,000,000,000 (twenty-five billion).

**TOMATO:** A medium-size tomato might weigh 150 g, so a tomato contains about 150,000,000,000 (150 billion) cells.

**BURGER:** A quarter-pound burger is about 115 g, but not all of this will be pure meat–it might also contain herbs, flour, or binding agents such as egg. High-quality burgers might be 95% meat, which means 115 x 95% = 109.25 g of the burger will be made up of meat cells. So there will be around 109,250,000,000 meat cells in the burger.

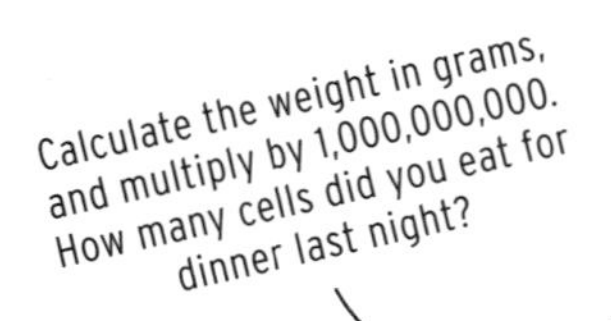

### OUR BODIES

The human body contains around 37.2 trillion cells—that's 37,200,000,000,000 cells. These are a mixture of different types that make up your skin, fat, muscle, blood, nerves, and many other parts of your body. But not all—some of your mass is made up of bones, hair, teeth, and nails, which are composed of proteins such as collagen and keratin. These parts aren't made of living cells, but they're produced by the cells of your body.

### HOW MANY IS THAT?

Large numbers like a billion, or a trillion, can be difficult to understand. Imagine if the cells were much bigger, and each cell was a single grain of rice. A 1-kg (2.2-lb.) bag of rice contains around 50,000 grains. To have the same number of grains of rice as there are cells in a tomato, how many bags of rice do you need? (The answer is on page 149.)

# SURPRISE CELLULOSE!

**So many of the objects we keep around us in our homes originate from plant-based materials, and anything that was originally part of a plant will be made up of plant cells, with tough cell walls made from cellulose. A lot of the things you see around you will contain cellulose!**

Can you identify which of the following household objects do and don't contain cellulose?

1. A pencil
2. Lettuce leaves
3. Ham
4. An apple
5. A $100 bill
6. Keys
7. A pair of jeans
8. Candy wrappers
9. Rocks
10. A nickel
11. A stick of celery
12. A knife and fork
13. A bell pepper
14. This book
15. A saucepan
16. A daffodil stem
17. A milkshake
18. A cotton T-shirt
19. A dinner plate

# **EXPERIMENT:** CHANGING THE COLOR OF FLOWERS

**Plants need water. They use it for photosynthesis (see page 33) and to keep their stems stiff and upright. If you stand a cut flower in water and wait a few hours, you'll see the water level drop.**

The way plants move water is through tubes of cells running all the way up through the plant, which are made of a tissue called xylem. Xylem has tough waterproof cell walls containing a substance called lignin, and the cells are arranged into long tubes. They transport water and minerals upward from the roots of a plant to the leaves and flowers using capillary action: water evaporates from the leaves—a process called transpiration—and more water is pulled in to take its place.

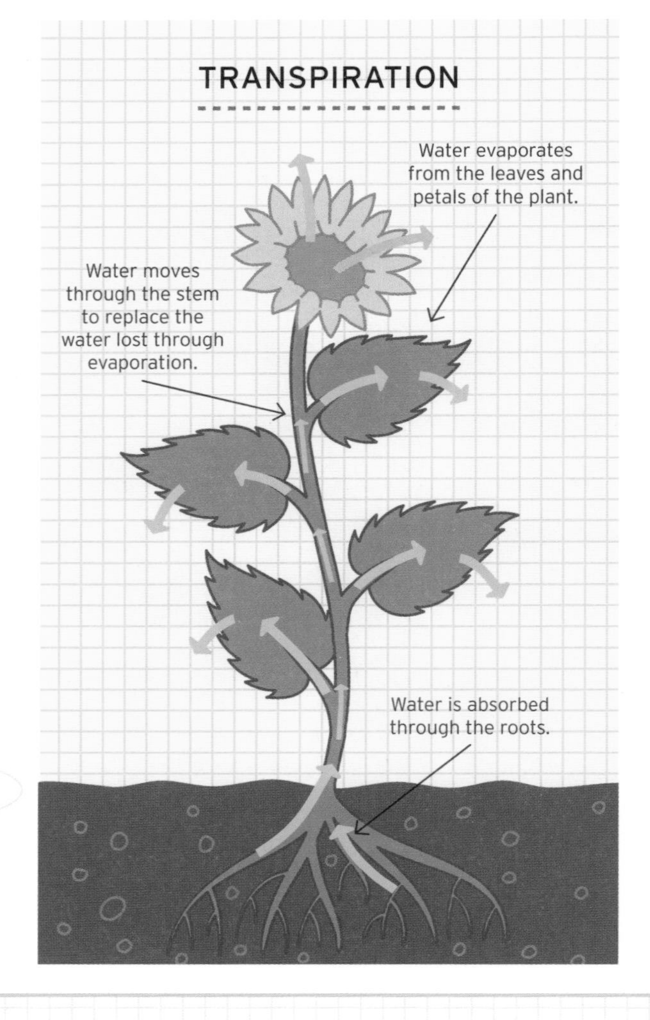

If you have some food coloring and a little patience, you can see water being transported up a flower stem.

### YOU WILL NEED:

- **White flowers (carnations and baby's breath work best)**
- **Clear glass vase, or tall drinking glass, for each flower**
- **Different colors of food coloring**

### WHAT TO DO:

**1.** Place each flower in a separate glass with a few inches of water. Add 10–20 drops of one color of food coloring to the water.

**2.** Wait 24 hours.

You should see that the colored water has been drawn up the flower stems and into the petals of the flowers. You might not be able to see the coloring in the stem or leaves, but the white petals will have changed color, or show color in some places.

## FUN FACTS

- Some colors work better than others and move up the plant more quickly. Try different ones to see which work best.
- Florists use this technique to create colorful bouquets. You could put some white flowers in one color of water for a while, then move them to another color. Try making some in the colors of your favorite flag!
- The food coloring won't harm the flowers (it's safe for humans to eat, and isn't toxic). It just sits inside the plant's cells and makes them look a different color.

# **DISCOVER:** WHY IS FRUIT TASTY?

**Many types of plants reproduce by creating seeds: tiny bundles of cells, enclosed in a tough coating, which can grow into new plants. Seeds range in size from tiny orchid seeds, which weigh 0.0001 mg, up to the world's largest known seed—a double coconut, which can weigh as much as 42 kg (92.5 lb.).**

Plants have a special trick for spreading their seeds—using tasty fruit.

If the seeds from a plant fall on the ground nearby and grow there, the plant might then find itself in competition with its offspring for resources—water, light, minerals, and space. It makes sense for a plant to want to reproduce far away from itself. But unless plants develop legs and the ability to walk, they won't be able to do this by themselves.

Some plants solve this problem by producing seeds that are mobile—dandelions, for instance, have fluffy white "parachutes" attached to their seeds, and sycamore trees have characteristic winged seeds, which twirl around like helicopter rotor blades and get carried away by the wind rather than landing on the ground nearby.

Other plants have seeds that are designed to be carried by water. Coconut trees, which are often found near water, grow seeds that have a tough, fibrous coating and float in water, and have been known to float long distances—up to 3,000 miles—before settling and growing on another island.

## SWEET TEMPTATION

Most flowering plants, however, have developed an ingenious solution that takes advantage of two things: the fact that animals exist, and the fact that sugar is tasty. Most seeds contain nutrients for the emerging plant to use once it starts growing, but the twist is that many plants grow a different type of nutritious material around the seeds, in the form of a fruit, which is high in sugars and brightly colored, and therefore attractive to foragers.

The idea is that animals (including humans) will be encouraged to eat the tasty fruit and then wander off elsewhere, later producing droppings that still contain some intact seeds. The seeds will find themselves in a new location, and they will already be in a pile of fertilizer! It's a clever way to take advantage of animals, and it's why fruit is a great source of nutrition.

### BREEDING WITHOUT SEEDING

Most fruit grown for human consumption is farmed, and the plants it grows on don't use seeds to reproduce. Farmers can create new fruit trees by taking cuttings from existing trees and growing them into separate plants, which means they grow similar fruit—naturally germinated seeds can mutate and show variation, and stores much prefer neat rows of identical apples! (Find out more about selective breeding on pages 140-141.)

# **DISCOVER:** FLOWERING PLANTS AND FRUIT

**Fruit grows on flowering plants and is produced inside the flower head—in the ovary, which is part of each flower. There are actually many categories and types of fruit, each with slightly different properties.**

## POLLINATION

Pollen is produced in a flower's anthers, brightly colored petals attract insects to spread the pollen, and sepals protect the petals. In the center of the flower are the reproductive organs, including a stigma and an ovary.

Once the flower has been fertilized—by pollen from another plant that has been deposited onto the stigma by an insect—the ovary is where the plant grows seeds and fruit.

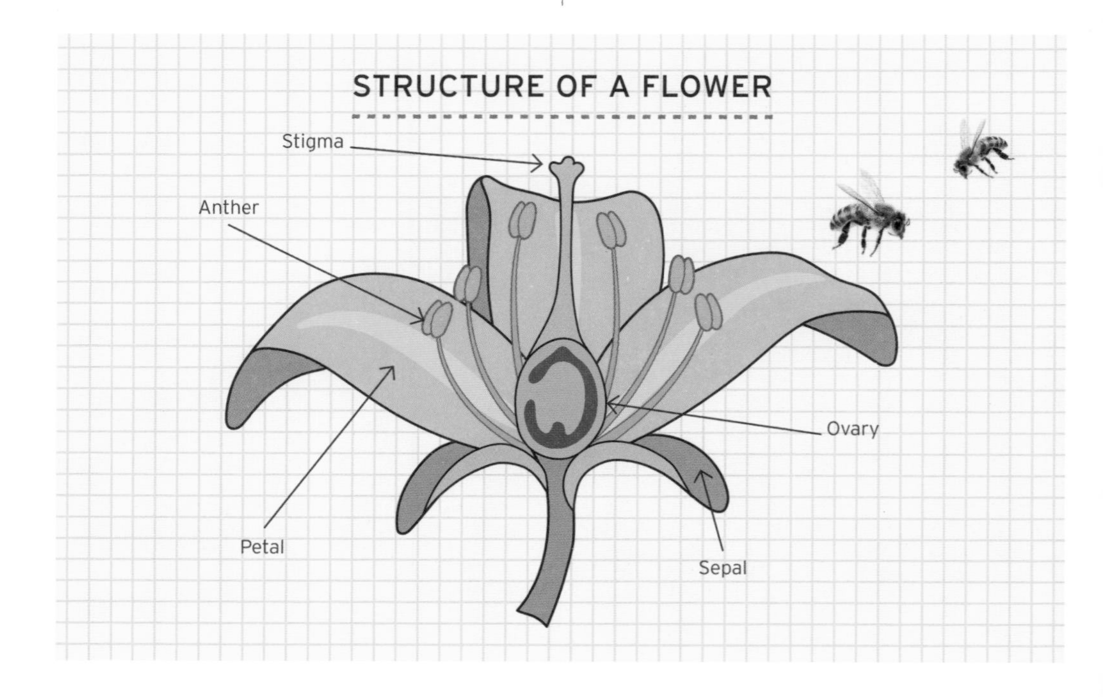

**TYPES OF FRUIT INCLUDE:**

- **True berries**

These are fleshy fruits without stones, produced from one flower with a single ovary.
**EXAMPLES:** Black currant, blueberry, cranberry, red currant, grape, kiwi, pomegranate, guava, tomato, eggplant, banana

- **Aggregate fruit**

These develop from a single flower with multiple ovaries, which join together as the flower grows, producing a fruit made up of multiple fleshy parts, each containing a seed.
**EXAMPLES:** Raspberry, blackberry, boysenberry

- **Hesperidia**

These are a modified form of berry with a tough, leathery rind and juicy interior that is split into segments. All citrus fruits are hesperidia.
**EXAMPLES:** Orange, lemon, lime, grapefruit

- **Pepos**

Pepos are another kind of modified berry, with a thick, hardened skin, sometimes called a rind. Unlike hesperidia, the interior isn't divided into segments.
**EXAMPLES:** Melon, pumpkin, squash, cucumber

- **Multiple fruit**

These are produced by a cluster of several flowers, whose fruits grow and join together to make a single fleshy mass.
**EXAMPLES:** Fig, pineapple, mulberry

- **Accessory fruit**

The flesh of accessory fruit partly grows from a different part of the flower, not the ovary.
**EXAMPLES:** Strawberry, apple

- **Drupes**

Also called stone fruits, drupes have an outer fleshy part containing a hardened shell with a seed inside.
**EXAMPLES:** Cherry, plum, apricot, peach, nectarine

- **Dry fruits**

These are fruits without a fleshy part. Some of them are dehiscent, meaning they burst open to spread their seeds.
**EXAMPLES:** Nuts, including beech, hazelnut, and acorn; legumes, such as peas, beans, and peanuts; and fibrous drupes, including coconuts and walnuts.

# **EXPERIMENT:** APPLES AND LEMON JUICE

**If you cut open an apple, it's white inside—but not for long! Apple flesh will turn brown if left out in the air. This experiment will show how dipping the apple in different liquids affects this.**

## YOU WILL NEED:

- **Apples of any variety, red or green**
- **Sharp knife to cut up the apples (ask an adult to help)**
- **Liquids: lemon juice, apple juice, water, white vinegar**
- **4 bowls, for dipping the apples in liquid**
- **5 plates**
- **Pair of tongs, or a fork for picking up apple pieces**

## WHAT TO DO:

**1.** Pour the liquids into four different bowls, and cut the apples into pieces. Aim for 15–20 pieces in total.

**2.** Use the tongs to dip 3–4 pieces of apple in one of the liquids, holding them under for 30 seconds, and then arrange them on a plate, labeled so you can remember which is which.

**3.** Repeat for the other liquids, cleaning the tongs between each set.

**4.** Set up a fifth plate with more pieces of apple, but don't dip them in liquid. This will be your control, which you can check against to see what effect the liquids have had.

**5.** Leave the plates for a few hours.

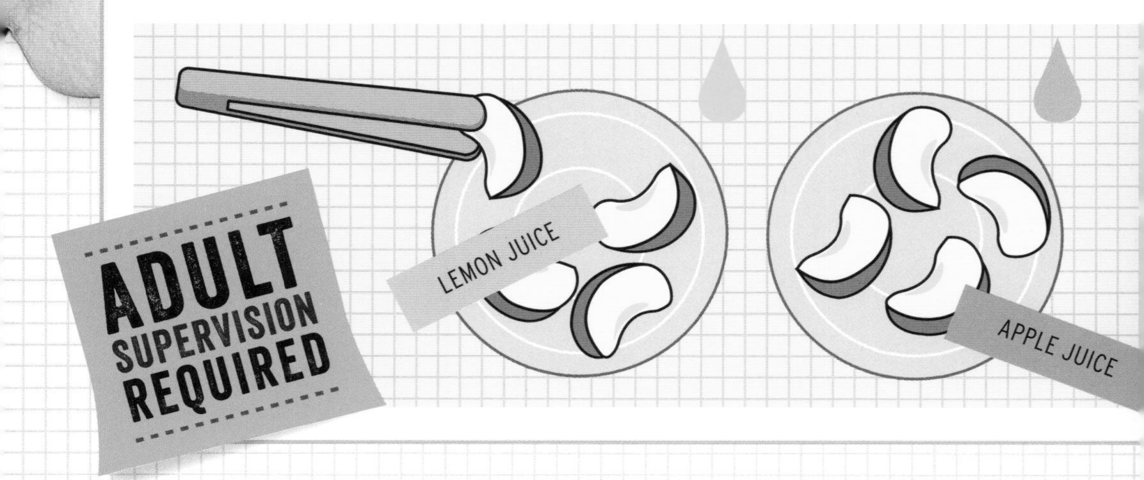

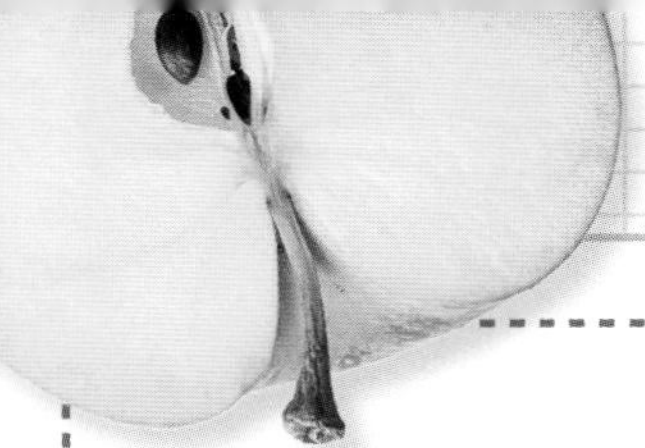

## WHAT HAPPENS?

### CONTROL APPLES (NO LIQUID)

The flesh of the apple pieces that weren't dipped in anything should turn brown. This could take anywhere from a few minutes to a few hours, depending on the type of apple. It happens because the apple contains an enzyme called polyphenol oxidase, which reacts with oxygen in the air to produce melanin (a brown pigment).

### LEMON JUICE

Lemon juice is strongly acidic and enzymes are proteins, which can be damaged by acidic environments. The lemon juice stops some of the enzyme from reacting with oxygen, so the apple will stay whiter.

### APPLE JUICE

Apple juice is also acidic, but not as strongly acidic as lemon juice. This means it will stop some browning, but not all (they won't be as brown as the control apples).

### WATER

Water isn't acidic, so it won't damage the enzyme, but it will soak into the apple and form a barrier, which will stop some oxygen from getting to the apple. The apple pieces will go brown, but less brown than the control apples.

### WHITE VINEGAR

Vinegar is also a strong acid, and will have a similar effect to the lemon juice, stopping the enzyme from causing browning (but you're less likely to enjoy the apples afterward!).

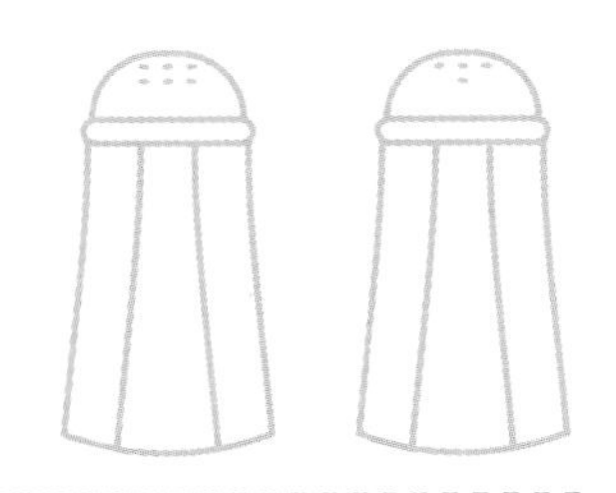

# DISCOVER: HOW DO PLANTS MAKE DINNER?

**All biological organisms need energy to live, and they get this energy from sugars, such as glucose. A process called respiration takes place inside cells, and produces the energy an organism needs to live.**

### RESPIRATION

The main type of respiration is called aerobic respiration, so called because it uses oxygen. The chemical reaction is:

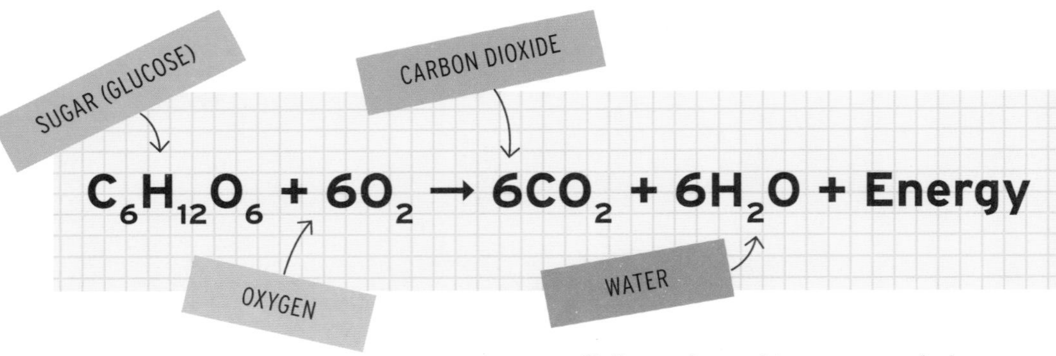


$$C_6H_{12}O_6 + 6O_2 \rightarrow 6CO_2 + 6H_2O + \text{Energy}$$

Cells are brought oxygen and glucose by the organism's transportation system—in humans and animals, this is the bloodstream. The respiring cells produce carbon dioxide and water as waste products, and the energy is released to allow cells to function.

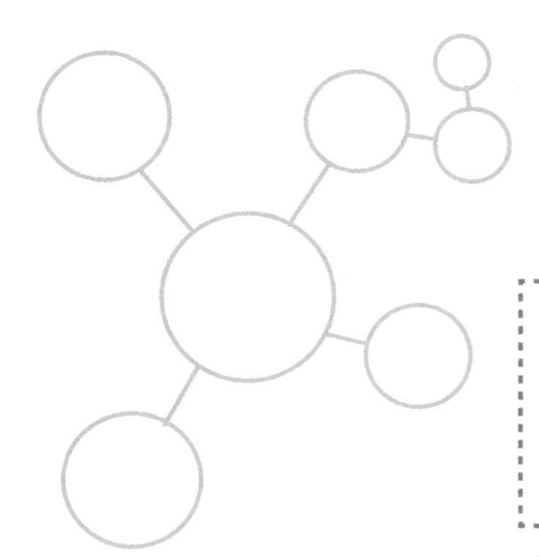

**RESPIRATION** happens in both plants and animals, but in plants there's a second process, which means they don't need to eat to get glucose molecules. They can produce their own sugar, using energy from the sun.

## PHOTOSYNTHESIS

Photosynthesis, which happens in all green plants, is the reverse process of respiration. It uses water and carbon dioxide, and produces glucose and oxygen. The chloroplasts in the cell (see pages 12–13) contain a green pigment called chlorophyll, which allows the plant to absorb light from the sun. The energy from sunlight allows the photosynthesis reaction to take place:

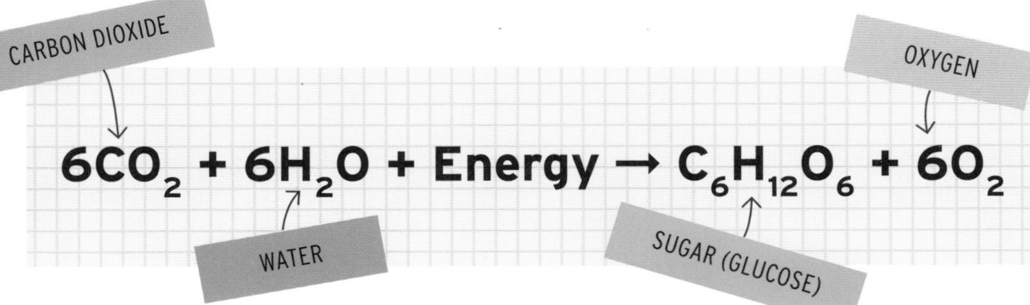


Oxygen is released from the plant into the air, and the glucose is used by the plant's cells for respiration. This is why plants need water and plenty of sunlight to live.

### SUNLIGHT OR LAMPLIGHT?

Plants need light to complete photosynthesis, and even though most plants live outside in the sunshine, they can still survive in a pot indoors (as long as you don't forget to water them!). Plants on the window ledge will survive better than ones in a room with no windows, so does this mean that sunlight is better?

Sunlight is different from artificial light. Most lamps don't emit as much energy in the red and blue region of the light spectrum as the sun does, and plants have evolved to use all the different wavelengths of light, so they like sunlight best. Light from the sun is also generally more intense than artificially generated light. Plus, it's free because it doesn't take any electricity to produce it!

# **EXPERIMENT:** DISSECTING AN ONION

**Onions are a staple in many different cuisines, and you will likely have eaten hundreds of them in your lifetime. Curry and stir-fry dishes, salads, soups, gravies and stews, hundreds of savory dishes and even chutneys and pickles all make use of onions' versatility and flavor.**

Anyone who cooks with onions will likely at some point have had cause to chop one up. The structure inside an onion can be seen as soon as you cut it open, and it has several interesting features. Dissecting an onion can provide an insight into how the onion was formed (it might also make you cry).

## YOU WILL NEED:

- **1 onion (red, white, or brown)**
- **Sharp knife (ask an adult for help)**
- **Cutting board**
- **Microscope and microscope slide, to look at onion cells (optional)**

## WHAT TO DO:

**1.** Use the knife to cut the onion in half, and look at the structure inside.

**2.** Peel apart two of the layers of onion from one of the halves. What do you notice? Rub your finger on the surface of the curved side of the onion. Is it rough or smooth?

An onion grows in layers. The layers are, biologically speaking, leaves, but in the onion plant they grow together around a stem to form a bulb. The tougher outside layers grow first, to protect the onion, and new layers grow from the inside.

Between each layer of onion flesh, there's a membrane, made up of a layer of onion cells. If you rub it, it will come away, and the onion left behind is smooth. If you have a microscope, you can peel some of the membrane off and put it onto a slide. Can you see the individual cells? They are about 100 micrometers across. Can you identify any structures within the cell?

## WHY DO ONIONS MAKE YOU CRY?

Onions don't smell strongly of anything when they're whole, but once you cut them open and damage the cells, enzymes are released that react with other chemicals present in the onion to produce syn-Propanethial-S-oxide. This is a gas that travels through the air and, when it contacts the nerve cells in your eyes, causes a stinging sensation and makes your eyes water. The same compound is also responsible for the heat and burning sensation you get when eating raw onions.

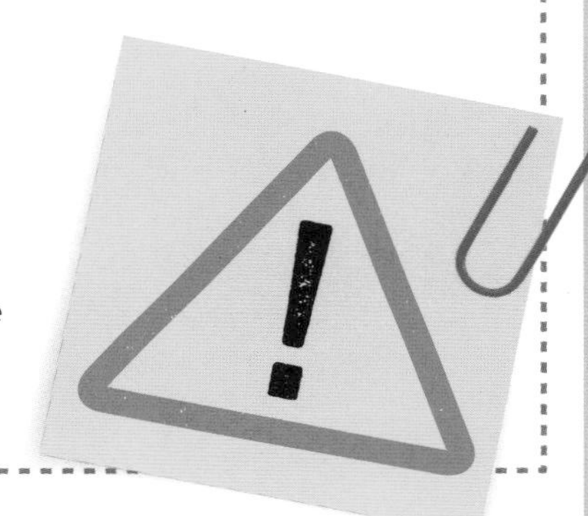

# **DISCOVER:** VEGETABLES

**Vegetables are an important part of our diet. They are a good source of fiber and essential vitamins and minerals, and they're mostly low in fat and calories. People who eat vegetables are at a reduced risk of cancer, stroke, and heart disease. Plus, they can be really tasty!**

The definition of "vegetables" is more culinary than botanical; some botanists consider there to be no such thing. The foods we call vegetables can come from many species and parts of plants.

## ROOTS

Carrots, turnips, and radishes grow as roots. The carrot plant itself is leafy, green, and looks like parsley, and the part you eat is actually the root, whose job is to absorb water from the soil.

## TUBERS

Potatoes and yams grow underground, on tuberous plants, which use specially thickened parts of their underground stem (or in the case of sweet potatoes, the root) to store energy and nutrients. So, when you dig up a potato, you're actually stealing the plant's food stash!

## BULBS

Some plants grow from bulbs, which also store energy for the growing plant, including onions, fennel, leeks, and garlic.

## LEAVES

Leafy vegetables are a great source of vitamins, minerals, and fiber, and include lettuce, kale, chard, and spinach. Nearly 1,000 different plant species are known to have edible leaves.

## STEMS

Asparagus, kohlrabi, and bamboo are all plant stems eaten as vegetables.

## FLOWERS

Some plants, including broccoli and cauliflower, have developed edible flowers. It might not look like it, but the part of broccoli you eat is actually made up of tiny flower heads!

## FRUIT

Plenty of vegetables are technically fruits, including tomatoes, eggplant, bell peppers, and squash.

## BUDS

Vegetables such as brussels sprouts and cabbage are actually the buds of the plant, and consist of multiple overlapping leaves in the shape of a ball.

## SEEDS

Peas and beans are types of seeds. Other seeds, such as mustard and rapeseeds, are also eaten as food. Seeds contain energy and nutrients for the growing plant, so they're often highly nutritious.

**BRASSICA** is a family of vegetables that spans many of these categories. It includes roots such as rutabaga and turnip, stems such as kohlrabi, leafy plants such as bok choy, kale, and collard greens, flowers such as cauliflower and broccoli, buds such as sprouts and cabbage, and seeds such as mustard seeds. The Brassica family is also sometimes called Cruciferae because the plants' flowers have four petals in a cross shape.

# **EXPERIMENT:** WHAT DO PLANTS NEED TO GROW?

**Plants have four basic needs that must be met in order to grow: soil, water, air, and light. But which of these are most important? This simple experiment will compare the effect of removing each of these basic needs one at a time, to see what effect it has on the growth of the plant.**

## YOU WILL NEED:

- **Plant seeds, such as radish, watercress, or mustard seeds**
- **7 small clear plastic cups or glasses**
- **Cotton balls**
- **Paper towel**
- **Soil (1–2 cups)**
- **Sand (3 cm [1 in.] in the bottom of a glass)**
- **Water**
- **Plastic wrap**

## WHAT TO DO:

Investigate the effect of changing the plant's environment by planting seeds in different conditions, and comparing their growth over a week.

## DO PLANTS NEED SOIL?

Set up four glasses: one with 3 cm of soil in the bottom (labeled "control"); one with sand; one with a few cotton balls; and one with a scrunched-up paper towel. Place a few seeds in each, on the surface (or just under it, if you're planting in soil or sand), and pour water in each: just enough to make the soil moist (and the same amount in each of the others). Cover the top of each cup with plastic wrap, and poke some holes in it—this will keep the moisture in, but not the air out.

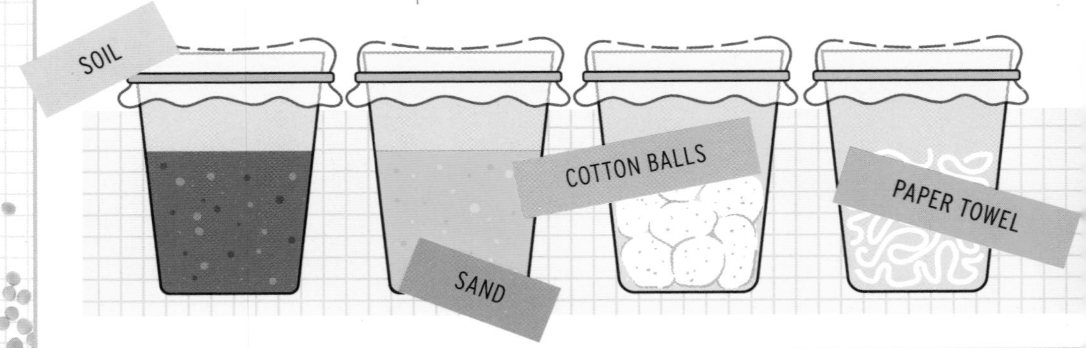

**Important:** Since in this experiment you're comparing the medium the seeds are grown in, you need to keep everything else the same. Make sure the plants are regularly watered through the week (only add water if the soil looks dry, but add the same amount to each) and that they're all left in sunlight, in a well-ventilated room.

## DO PLANTS NEED AIR?

Set up another cup with soil, seeds, and water, and cover it with plastic wrap, but don't poke any holes in it. You won't be able to take the cover off to add water without letting air in, so make sure it's well watered to start with (but not too much!). You can compare this with the "control" from the first group, so place it next to the first four—meaning it will get light.

## DO PLANTS NEED WATER?

Set up a sixth cup with seeds in soil, and cover it with plastic wrap with holes poked in it, but don't add any water. Place it alongside the other five. You can compare this with the control, which is getting watered.

## DO PLANTS NEED LIGHT?

Set up a seventh cup with soil, seeds, and water, and cover it with plastic wrap with holes poked in it. Place this cup in a dark but well-ventilated place (such as inside a cupboard). Don't forget to water it!

## WHAT HAPPENS?

Seeds will germinate happily without soil, so small seedlings should appear in all of the cups testing growth mediums, but without air, water, or light the plants won't grow well, if at all.

What happens if you leave them for longer than a week? Do the seedlings grow differently in the different mediums?

# **DISCOVER:** RECORD-BREAKING VEGETABLES

**The experiment on pages 38–39 proved that to germinate, seeds need water, light, and air, though they will start growing without soil. However, once the plant grows beyond a small seedling, it needs more nutrition, and a well-balanced soil containing nutrients and minerals is necessary for its continued growth.**

Soil naturally contains nutrients like nitrogen, phosphorous, potassium, and calcium. When these nutrients aren't available, plants suffer from nutrient deficiency and don't grow properly. Plants not getting their proper supply of nutrients can show changes in color, slower growth, patches on the leaves, cracks in the stem, and rotting or dimpled fruit. Fruits and vegetables are particularly affected by this, so farmers need to be careful to protect their crops.

**RECORD-BREAKING VEGGIES:** which of these are taller, or heavier, than you?

**HEAVIEST CARROT:** 10.17 kg (22.44 lb.), grown by Christopher Qualley (USA) in 2017

**LONGEST CARROT:** 6.25 m (20 ft., 5.86 in.), grown by Joe Atherton (UK) in 2016
*Including its roots, it was as tall as a house!*

**HEAVIEST TOMATO:** 3.906 kg (8.61 lb.), grown by Dan Sutherland (USA) in 1986
*About the same weight as 285 empty soda cans!*

**HEAVIEST PUMPKIN:** 1,190.23 kg (2,624 lb.), grown by Mathias Willemijns (Belgium) in 2016
*That's heavier than a bull!*

Gardeners and farmers use fertilizers, made from organic matter or chemicals, to increase the nutrient levels in the soil. Growing crops repeatedly in the same soil can lead to reduced levels of minerals (see page 132 for one way to combat this), as can heavy rain and flooding, which can wash away nutrients.

**LONGEST CUCUMBER:** 107 cm (42.1 in.), grown by Ian Neale (UK) in 2011

**HEAVIEST CABBAGE:** 62.71 kg (138.25 lb.), grown by Scott A. Robb (USA) in 2012

## GIANT VEGGIES

If you want to find botanical fame and fortune, one way is to grow giant vegetables. By carefully choosing the right varieties of seeds, and manually pollinating the plants, you can choose which plants breed, and grow bigger and bigger veggies. You'll also need to give your plants plenty of light, water, and air, a good-quality fertilized soil that's nicely aerated (containing lots of air gaps, to allow water and nutrients to get to the roots), and plenty of room to grow.

# **EXPERIMENT:** MUSHROOM SPORE PRINTS

**Mushrooms are considered a vegetable, but they're very different than most other plants we eat. Mushrooms are the fruiting bodies of fungi, which means they're created by the fungi in order to spread their spores, which they use to reproduce.**

The spores are tiny, consisting of a single cell and forming a powder-like substance. This drops from the underside of the mushroom, and is carried away by water evaporating from the mushroom and by the wind, so it can land elsewhere and grow into a new plant. Scientists use this natural property of the spores to make a picture, called a spore print, which they can use to identify mushrooms based on the shape and color of the resulting image.

## YOU WILL NEED:

- **Large, mature flat mushrooms (ask an adult for help)**
- **Blank white paper**
- **Hair spray, or other spray fixative (optional)**

## WHAT TO DO:

**1.** The underside of the mushroom should have visible black spores. You might need to remove the stem to allow it to lie flat.

**2.** Place the mushroom, spore side down, on a piece of paper, and cover it with a glass or bowl to make sure it isn't disturbed.

**3.** Leave it in place for a few hours, or overnight, if possible.

**4.** Remove the bowl and carefully lift off the mushroom. Tiny black spores falling from the underside will have created an image of the mushroom on the paper. Don't smudge the image!

**5.** You could use hair spray or other spray fixative (sprayed from far away, so you don't disturb the spores) to set the picture, so it can be kept.

### SOME MUSHROOMS ARE POISONOUS!

Mushrooms have been harvested and eaten for centuries, but you can't just eat any mushrooms you find lying around. Some varieties of mushroom are incredibly poisonous, and eating them can lead to sickness, hallucinations, and even death. If you're going to eat mushrooms you didn't buy in a grocery store, make sure you've got someone with you who knows which ones are safe to eat!

# **DISCOVER:** POTATOES

**Potatoes are an important food source, and the number-one vegetable crop in the world. In 2017, 388 million tonnes (428 million tons) of potatoes were produced globally. Find out all about them here:**

Above ground, potato plants have leaves and flowers.

- Potatoes are a tuber, a part of a plant that's been specially adapted to store food and nutrients, made from an underground section of the stem that grows into a round lump.
- Above ground, the plants potatoes are harvested from are similar to tomato plants. They are from the nightshade family, which includes eggplant and tobacco. If their flowers grow and are pollinated, they produce a small, green, inedible (toxic) fruit similar in appearance to a tomato.

• Tubers are filled with starch molecules—and some sugar—for the plant to use during winter, when shorter days mean there's less sunlight.

• There are many varieties of potato, including red, yellow, white, and even purple. For cooking purposes, we distinguish between floury potatoes, which contain more starch and are good for baking and mashing, and waxy potatoes, which are better for boiling.

• Potatoes are used to make fries, chips, hash browns, and tater tots; they're eaten boiled, roasted, baked, and mashed, and as an ingredient in stews, salads, and casseroles. Potato can also be used as a thickener or binder in processed foods.

• Even though potatoes are relatively low in fats and sugars, and the skin and outer flesh contain lots of fiber, many potato dishes involve frying in oil, or adding butter or cream, which means they may not be so healthy. But when baked or boiled and served as a side dish, they're pretty good for you! They're a good source of healthy antioxidants, vitamins, and minerals.

## THE POTATOES HAVE EYES!

On the surface of a potato, you'll find "eyes"—the parts which grow into new potato plants. If you leave your potatoes in a dark cupboard for slightly too long, they'll think they've been planted and might start to sprout! To avoid this, keep them cool (under 5ºC [40ºF]) and dry. Store-bought potatoes are specially treated to prevent this from happening, but if they do sprout, they're still safe to eat. As long as they're still firm to the touch and not too shriveled, just trim off the sprouting parts and any soft spots and you're good to go.

# LEARN ABOUT: PHOTOSYNTHESIS

**Respiration and photosynthesis are opposite processes, creating and using up oxygen and carbon dioxide. Every living creature respires, but only plants photosynthesize (along with some species of bacteria and algae). While most creatures on Earth's surface take in oxygen and release carbon dioxide, plants also do the opposite.**

Test your photosynthesis knowledge with this true-or-false quiz!

## TRUE OR FALSE

**1.** On average, plants produce more carbon dioxide than they take in.

**2.** One human produces as much carbon dioxide as can be taken in by about eight trees.

**3.** Having plants around the house makes the air cleaner.

**4.** A leaf gives off about five milliliters of oxygen per hour.

**5.** When tree leaves turn orange in the fall, it's because they've stopped producing chlorophyll.

**6.** Plants can photosynthesize just as well in the dark.

**7.** Plants and some bacteria and algae are the only living things that create food by photosynthesis.

**8.** Carbon dioxide passes into the leaves through pores in the surface of each leaf.

**9.** Plants with red leaves don't contain any chlorophyll.

**10.** An area of rain forest the size of South Carolina is cut down every year.

## WHAT IS THE GREENHOUSE EFFECT?

Carbon dioxide and other gases, when released into the atmosphere, form a layer surrounding Earth. This traps heat which would otherwise be able to get out, just like a greenhouse does, and increases the temperature of the planet. Check out pages 126–127 to find out more.

# **LEARN ABOUT:** POTATOES

**The average American eats around 64 kg (140 lb.) of potatoes per year. But how well do you know the fourth most important staple crop? Try our potato quiz to see if you can remember, estimate, work out or guess the answers.**

## POP QUIZ: POTATOES

**1.** What part of a plant is a potato?
a) A stem
b) A fruit
c) A root
d) A seed

**2.** Potato is a member of the family Solanaceae. Which of the following is not a member of this family?
a) Tobacco
b) Nightshade
c) Tomato
d) Sweet potato

**3.** What is the meaning of the Latin word "tuber," from which tubers get their name?
a) Below, lower, farther down
b) Lump, bump, swelling
c) Earth, land, soil
d) Store, stock, storeroom

**4.** What proportion of potato flesh is water?
a) 59%
b) 69%
c) 79%
d) 89%

**5.** How many potatoes grow at one time on the average potato plant?
a) 2-5 potatoes
b) 5-15 potatoes
c) 15-25 potatoes
d) 25-30 potatoes

**6.** How heavy is the largest potato ever grown?
a) 900 g (2 lb.; the weight of a bag of sugar)
b) 2.49 kg (5 lb., 6 oz.; the weight of 17 bananas)
c) 4.98 kg (10 lb., 14 oz.: the weight of 8 basketballs)
d) 5.44 kg (12 lb.; the weight of 54 blueberry muffins)

**7.** When are main crop potatoes harvested?
a) July
b) August
c) September
d) October

**8.** Roughly how many different varieties of potatoes are there?
a) 200
b) 400
c) 2000
d) 4000

**9.** Potatoes were originally domesticated in 8000-5000 BCE—but where?
a) South America
b) Africa
c) North America
d) Europe

**10.** How many potatoes would you need to eat to get the same amount of energy as in a regular 100 g (3½ oz.) candy bar?
a) 2-3
b) 6-7
c) 8-10
d) 12-15

# DISCOVER: HERBS AND SPICES

**One way you'll often see plants used in the kitchen is in the form of herbs and spices—foods with a strong flavor, used in small amounts to give a dish a particular taste.**

## HERBS

Botanically, herbaceous plants are ones that grow green, leafy shoots with no woody stem, and whose above-ground parts die back after each growth season. Herbs can be annuals, biennials, or perennials.

- **Annual:** A plant whose life cycle, including germination and production of seeds, lasts one year. At the end of the year it dies, leaving behind seeds to grow into a new plant.
- **Biennial:** A plant whose life cycle lasts two years, with one dormant period during the colder months.
- **Perennial:** A plant that lives longer than two years, dying back in the fall and returning in spring. Their roots survive underground.

Herbs contain chemical compounds that give them their strong flavors and smells. These are released when you bruise the leaves, by crushing or chopping. Many herbs don't smell much before you rub one of their leaves between your fingers, then the cells are broken open, and the chemicals are released into the air, so you can smell them.

### DO YOU LIKE CILANTRO?

Cilantro, a green leafy herb used widely in Latin American and Middle Eastern cooking, has a distinctive flavor, but some people find it more distinctive than others. In some parts of the world, due to a genetic variant, up to 20% of people taste cilantro as a horrible soapy flavor. The parts of the world in which this is least common (2–3% of the population) are Latin America and the Middle East!

## SPICES

Spices are not living plants but substances extracted from different parts of plants. These include dried, crushed, and powdered plant seeds and pods, roots, and bark. Spices can have warm flavors, and make foods spicy or aromatic.

## WHY ARE CHILIS HOT?

Chili peppers and other spicy foods contain a molecule called capsaicin, which stimulates the parts of the skin and tongue that normally sense heat and pain. This is a plant's defense mechanism to stop certain types of animals from eating their fruit—but it turns out humans enjoy hot sauce too much to let it put them off!

Different varieties of chilis, and different parts of the fruit, have different amounts of capsaicin—the seeds and the white flesh are usually the strongest. In general, bigger, green chilis tend to be less spicy, while smaller, red ones are more likely to give you a painful mouthful. Hundreds of varieties exist, including red, green, orange, yellow, purple, and even black-colored chilis, in sizes from 1 cm (¼ in.) long to more than 30 cm (1 ft.).

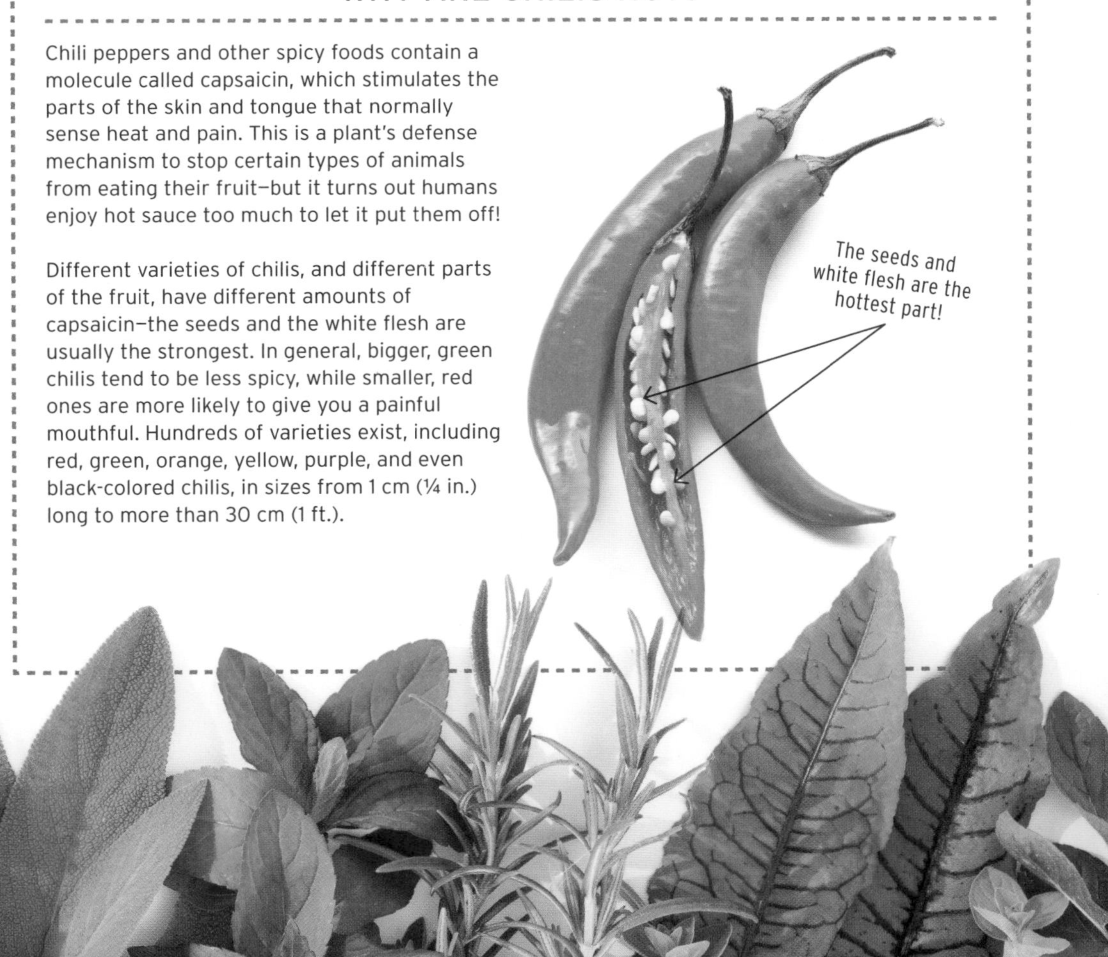

# CHAPTER 2
# FOOD

DISCOVER...

LEARN...

EXPERIMENT...

# DISCOVER: YEAST

**Yeast is a single-celled organism, meaning each individual cell of yeast is a separate entity. It's a fungus, and its cells (as we saw on page 13) are much like animal cells, with a nucleus, cytoplasm, ribosomes, and mitochondria, but also with a cell wall to hold it together.**

Yeast is used in making various foods, including many types of bread and risen cakes and pastries, as well as in vinegar. The yeast cells ferment carbohydrates such as sugars, and produce carbon dioxide. The Latin name for baker's yeast is *Saccharomyces cerevisiae*, which means "sugar-eating fungus."

This fermentation process is part of the organism's natural life cycle. Yeast, along with some sugar for it to eat, is added to a bread or pastry dough. The resulting carbon dioxide forms bubbles inside the dough, which make it rise.

There are over 500 species of yeast, and different varieties survive better in different environments, depending on the temperature and concentrations of toxins in their environment. Yeast is often dried, and sold as a powder—this doesn't mean it's dead, but it remains dormant until warm water is added to activate the yeast cells.

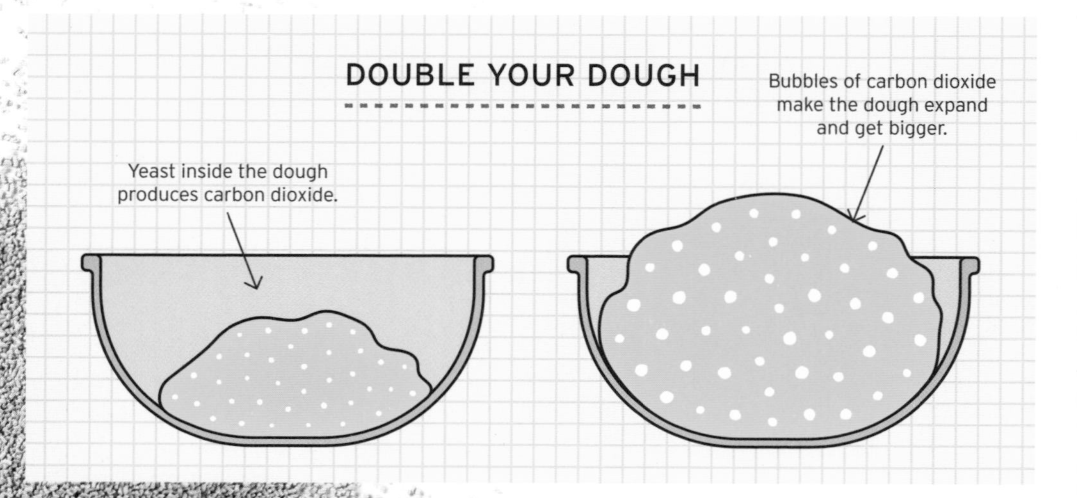

### YOU RAISE ME UP

Modern cakes use baking soda to create bubbles of carbon dioxide, which make the cake rise. Baking soda is the common name for sodium bicarbonate, and the heat applied during baking causes it to react and break down, releasing carbon dioxide as a gas. Baking soda can be used along with acidic ingredients like buttermilk, lemon juice, or cocoa to get more rise.

Historically, however, baking soda wasn't used as a raising agent until the 1840s. Prior to that, bakers used yeast to create risen cakes and bread, by feeding the yeast sugar to ferment and digest. Traditional sourdough bread uses a mixture of bacteria (*lactobacilli*) and yeast in a similar way.

As well as risen breads, plenty of modern cakes and baked goods still use yeast in their recipes, including donuts, cinnamon buns, and crumpets.

## YEAST EXTRACT

Yeast extract is made from dead yeast cells, whose cell walls have been destroyed by adding salt and heating. The enzymes released break down the proteins found in the yeast cells, resulting in a substance with a strong savory flavor, which can be used as a flavoring in stocks, sauces, soups, and gravies. It's the reason barbecue-flavor potato chips taste the way they do!

In the UK, Germany, and Australia, yeast extract is sold as a dark-brown spread (Marmite, Vegemite), and it's also sometimes used to make hot drinks.

# **EXPERIMENT:** YEAST AND BALLOONS

**Under the right conditions, yeast can produce carbon dioxide, and this is exploited by bakers to make bread and other baked goods. This experiment investigates exactly what conditions yeast needs to produce the most carbon dioxide.**

By placing a balloon over the neck of a plastic bottle, you can collect all the carbon dioxide that's generated by the yeast. Then, by comparing the sizes of different balloons with yeast under different conditions, you can see which conditions lead to more or less carbon dioxide.

### YOU WILL NEED:

- **6 identical plastic bottles of any size**
- **6 balloons**
- **Yeast**
- **Salt**
- **Sugar**
- **Water**
- **Funnel**

### WHAT TO DO:

Set up the six bottles with slightly different contents as shown below, and quickly stretch a balloon over the neck of each bottle once you've added the ingredients. Observe what happens over the next 5–10 minutes.

## WHAT HAPPENS?

The yeast in bottle 1 should produce the most carbon dioxide, and the balloon should inflate the most. Warm conditions, with sugar to digest, are the ideal conditions for yeast. Bottle 2, with cold water, might produce some carbon dioxide, but not as much as bottle 1. The other bottles—with salt, and without sugar—won't produce much carbon dioxide.

In factories where yeast is used to produce foods and drinks, the temperatures, and ratios of sugar and yeast, must be carefully monitored for optimal fermentation. Depending on the species of yeast used, the optimal temperature would be in the range of 15–35°C (59–95°F). Bakers who use yeast in breadmaking include an extra step called proofing, which gives the yeast time to ferment in a warm atmosphere.

## A GOOD EXPERIMENT

Varying the conditions in this way—keeping all the other factors the same, and varying one thing each time—is part of designing a good experiment. If you changed two things, and observed a change in the result, you wouldn't know which of the two things you'd changed caused the change in the result.

# **DISCOVER:** DIFFUSION AND OSMOSIS

**There are two important processes that take place in all living things, which move things around inside cells and organisms. One is diffusion, and the other is osmosis.**

## DIFFUSION

You probably know that if you put a drop of ink into a glass of water, it'll spread out. This is called diffusion, and it happens when you have a difference in concentration. In the case of the drop of ink, the ink molecules are in high concentration within the drop of ink, and in low concentration in the rest of the water.

As the molecules move around, the concentrations become more equal: molecules move from areas of high concentration to areas of low concentration to equalize. After enough time, the glass of water will have ink spread through it evenly, with the same concentration everywhere.

Diffusion occurs inside living organisms, such as in the intestines, where molecules of nutrients diffuse through the intestinal wall into the bloodstream. (See pages 84–86 for more about how your food is absorbed.) Another example is in the lungs: when the concentration of oxygen is high in the lungs and low in the blood, it will diffuse through into your bloodstream.

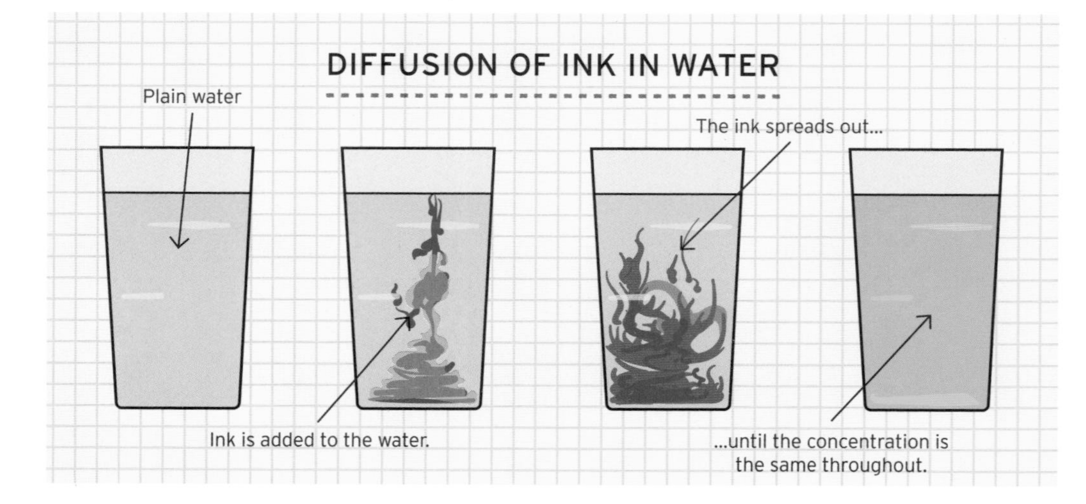

## CONCENTRATION

When a substance is dissolved in a liquid such as water, the concentration is the amount of that substance present for a given volume of liquid. For example, if you put 20 g of salt into a liter of water, the concentration would be 20 g per liter.

If you know the mass of the substance and the mass of the liquid, you can work out the concentration as a percentage. If you added 20 g of salt to 100 g of water, you'd have a 20-percent solution, calculated as:

$$\frac{\textbf{Mass of dissolved substance}}{\textbf{Mass of liquid}} \times \textbf{100} = \textbf{concentration (\%)}$$

You can also work out the concentration of gases in air, and of different liquids in a mixture.

## OSMOSIS

Osmosis occurs when a membrane—called a semipermeable membrane—is thin enough to allow water through, but stops larger molecules.

It works the opposite way to diffusion; when the liquid on one side of the membrane has a high concentration of something dissolved in it, and a low concentration on the other side, osmosis will work by moving water across to make the concentrations of the dissolved substances equal.

This means water will move from where the concentration is low to where it's high, so that the concentrations eventually become equal (and given enough time, they'll become exactly equal).

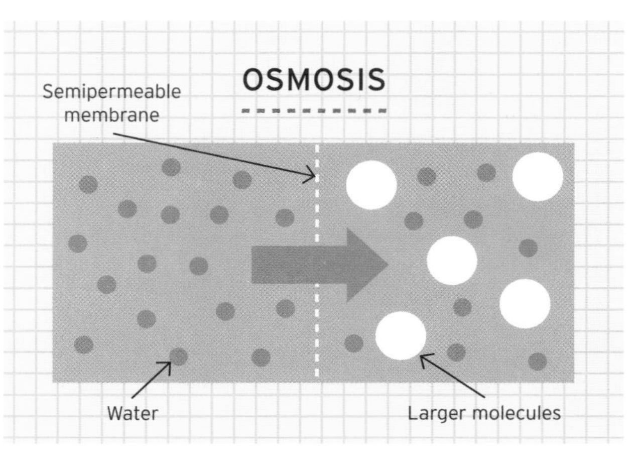

# **EXPERIMENT:** POTATOES AND OSMOSIS

**The cell membranes of plant and animal cells are semipermeable—water can pass through, but molecules dissolved in the water, like starches and sugars, can't. This leads to osmosis, and you can investigate how this works using potato cells.**

## YOU WILL NEED:

- Large potato
- Water, and several glasses or pitchers
- Salt
- Sugar
- Kitchen scale and measuring cups
- Ruler
- Pens and labels
- Sharp knife (ask an adult to help)

## WHAT TO DO:

**1.** Cut one end off the potato, and then cut off three 5-mm (1/4 in.) slices. Stack the circular slices together and trim them all to roughly the size of the smallest one.

Then cut a 10-mm (1/2 in.) slice, trim it to a 20-mm (1 in.) square, and cut it in quarters to obtain four 10-mm cubes of potato.

**2.** Set up three glasses to soak the slices. Pour 200 ml (7 fl. oz.) of water into each, then add 20 g (4 tsp.) salt to one and 20 g sugar to another. Leave the third as plain water, and put one potato slice in each glass. Leave them in for half an hour.

**3.** For the cubes, make up the following concentrations of salt in water: 10%, 5%, 1%, and 0% (plain water). One gram (1/5 tsp.) of salt in 100 ml (3 1/2 fl. oz.) of water gives a 1-percent solution, so you'll need the following quantities:
10% solution: 200 ml with 20 g salt
5% solution: 200 ml, with 10 g (2 tsp.) salt
1% solution: 200 ml, with 2 g (1/2 tsp.) salt
Plain water: 200 ml

**4.** Before adding the cubes to the water, make a note of their sizes (in case they're slightly more or less than 10 mm) and weights. Add one cube to each solution and leave it for half an hour.

SETTING OUT YOUR EXPERIMENT
The three slices
Cubing the potato
STEP 1
Soaking the slices
Sugar and salt
STEP 2
Making up solutions
STEP 3
Soaking the cubes
STEP 4

## WHAT HAPPENS?

**In the first part of the experiment, the flat round pieces of potato were placed in different solutions. Potato cells contain starch and sugar molecules dissolved in water, so the concentrations inside and outside the potato would have been as follows:**

| | PLAIN WATER | STRONG SALT SOLUTION | STRONG SUGAR SOLUTION |
|---|---|---|---|
| INSIDE POTATO CELLS | Some sugar | No salt | Some sugar |
| OUTSIDE POTATO CELLS | Plain water | Lots of salt | Lots of sugar |

Osmosis causes water to move into or out of the potato, so that the concentrations of each substance in the water are the same on both sides.

• In the salty water, the concentration of salt is high outside the potato, so water would move out of the potato cells to dilute the solution outside. This piece of potato should become softer and bendable, as water has moved out of it.

• In the sugary water, there's still a higher concentration of sugar outside, but the difference is less, as there's some sugar inside the potato. Water will still move out of the potato cells, but not as much, so it'll be bendable, but not as much as the first one.

• In plain water, the concentration of sugar is higher inside the potato, so water should move in to equalize the concentration. This should make the piece of potato more rigid and less bendable.

In the second part of the experiment, the cubes were weighed and measured before being put in the solutions, which were different strengths of salt solution.

- In the plain water, the concentration of sugars is higher inside the potato cells, so water should move into the potato, to make the concentrations more equal. Its size and weight should have increased slightly.

- In the weaker salt solutions, the concentration of salt outside the potato is only slightly higher, so some water will move out (as well as water moving in, to balance the sugar gradient!). It should be about the same size and weight.

- In the stronger salt solutions, there's a much higher concentration of salt outside, so water will move out to dilute it and make the concentrations more equal. The piece of potato should reduce in size and weight slightly.

It might seem that soaking a piece of potato in a liquid would cause it to soak up water and get bigger, but if the concentration outside the potato is higher, it'll cause osmosis to draw water out of the potato.

## DISCOVER MORE

Osmosis is one of the most important biological processes, and cells in plants and animals make use of it to regulate the movement of water around organisms. Plants absorb water through their roots, and it evaporates by transpiration from their leaves, which draws water up the plant and supplies all its cells with water. Animal cells, which don't have a cell wall, can shrink if they lose water, or burst if they take on too much.

# DISCOVER: BACTERIA AND MOLD

**Bacteria are single-celled organisms, and mold is a type of fungus—made of the same kind of cells as yeast, but with the cells joined up into long filaments. Both can be found in and around food, and only sometimes is it a bad thing.**

## THE GOOD

Bacteria, in combination with yeasts and mold, have been used for thousands of years in the manufacture of fermented foods like cheese, pickles, soy sauce, sauerkraut, vinegar, and yogurt. Much like yeast, bacteria consume sugars and produce other substances, creating unique flavors and textures in the food.

Yogurt is produced by bacteria that feed on lactose, the sugar present in milk. They convert it into lactic acid, which gives yogurt its tangy taste, and reacts with the milk proteins to give yogurt its thick texture.

Cheese is made using a similar process, and different types of bacteria are used to make different kinds of cheese. Some produce only lactic acid, giving a clean, acid flavor to cheeses such as cheddar. Other cheeses, like Emmental, are made using bacteria that also produce carbon dioxide when they ferment the milk, and the resulting bubbles give the cheese its trademark holes.

This is the point where mold is used in cheesemaking: spores of mold can be added to ripen the cheese. In cheeses like Camembert, the ripening occurs on the outside, giving the cheese a harder rind on the outside and a softer middle; for some blue cheeses like Stilton or Gorgonzola, the mold is mixed into the cheese for it to ripen throughout.

The mold breaks down the fat and protein in the cheese, producing chemicals which give it a stronger, sharper flavor. Without ripening, cheese can taste bland, and all strongly flavored cheeses have been through some kind of ripening process.

## THE BAD AND THE UGLY

Of course, mold and bacteria on your food isn't always a good thing; if you're not careful, bacteria getting into your mouth and digestive system can lead to some pretty nasty symptoms. Mold grows on food if it's not stored properly, and some species of mold produce toxins that will make you sick if you eat them. (See pages 72-75 for more on the bad side of bacteria and mold, and how you can stop it!)

Not the kind of mold you want to eat!

PARMESAN

# EXPERIMENT: GROWING MOLD

**Mold is present all around us, but it can be too small to be seen, and most varieties of mold are harmless. It's a fungus which grows on food when left for enough time in the right conditions, producing a visible fuzzy growth.**

If mold is visible, it's because it has formed a colony, consisting of many cells which form a single organism (a mycelium). The mold digests the sugars and starches found in the food in order to grow, and produces long filaments of cells which go all the way through whatever it's growing on.

Like all fungi, mold reproduces by releasing spores into the air, which can land on anything left open to the air and start to grow. Mold growing on food can produce toxins that make the food unsafe to eat, and some varieties of mold can cause a reaction in people with allergies.

Mold grows best in damp places with poor airflow. This experiment shows what happens to food placed in a plastic bag with a little water and left for a while. Although there will be spores of mold present on the bread already, some household dust will be added to help it grow.

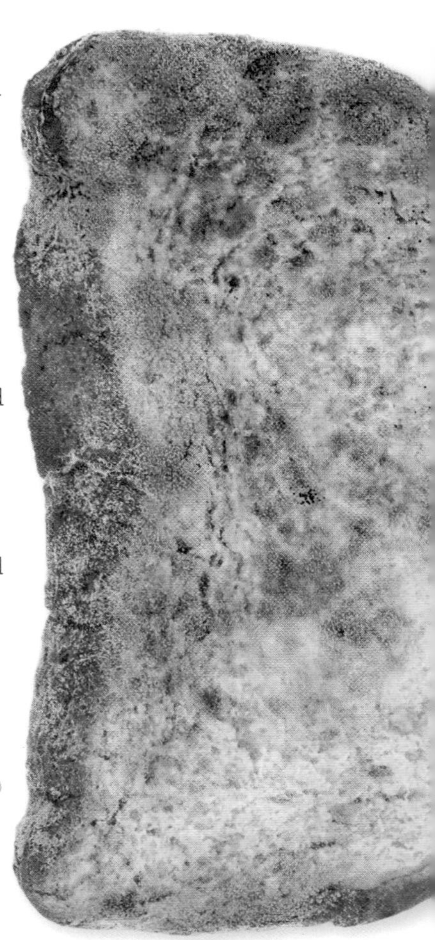

## YOU WILL NEED:

- 1 slice of bread
- Ziplock plastic bag
- Cotton swab
- Water
- Cardboard box, or a clean used carton

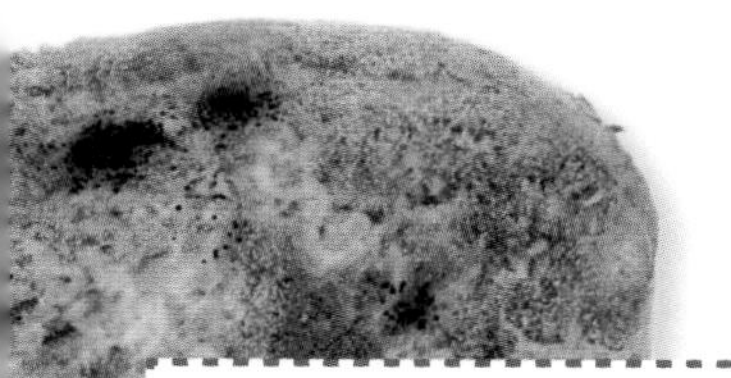

## WHAT TO DO:

**1.** Use the cotton swab to pick up some household dust from a surface that hasn't been cleaned recently, and rub it onto the center of the piece of bread.

**2.** Add a few drops of water to the middle of the slice of bread, and place it in the plastic bag. Seal the bag.

**3.** Put the sealed bag inside the box or carton, so it's in the dark and there is no airflow.

**4.** Leave the mold to grow for a few days—check the mold a couple of times a day to see how it's growing.

## WHAT HAPPENS?

It usually takes around 7–10 days for mold to grow on bread, but since you've kept the bread sealed and moist, and deliberately added some dust, it may start showing signs of mold sooner. Bread mold can be white, green, blue, or black, and mold also grows on fruits and vegetables. You should be able to see visible fuzzy or colored patches on the bread, and you may also be able to detect a strong smell.

## INVESTIGATE

Extend this experiment! Remember to always change one variable at a time, and maintain a control sample. Can you think of anything else?

- Does mold grow faster on stale or fresh bread?
- Does mold grow better in a cold or warm place?
- Does mold grow slower if you don't add water?
- Will mold grow on toast?

# EXPERIMENT: CULTIVATING BACTERIA

**This experiment lets some bacteria grow, so you can see what it looks like—bacteria are far too small to see without using a microscope, so you'll need to culture a colony of bacteria.**

Bacteria are all around us in small quantities, and they can be dangerous if allowed to grow. That's why it's important to wash your hands before eating, and to clean food preparation surfaces regularly. But here the aim is the opposite: to produce some interesting bacteria!

## YOU WILL NEED:

- 250 ml (8½ fl. oz.) water
- 2 ¼-oz. packages of gelatin (or agar)
- 1 package of low-sodium beef broth, or 2 g (½ tsp.) yeast extract
- 4 g (1 tsp.) sugar
- Large heatproof pitcher
- 2–4 shallow dishes (ideally glass)
- Cotton swabs
- Plastic wrap
- Oven mitts (for handling hot liquids)

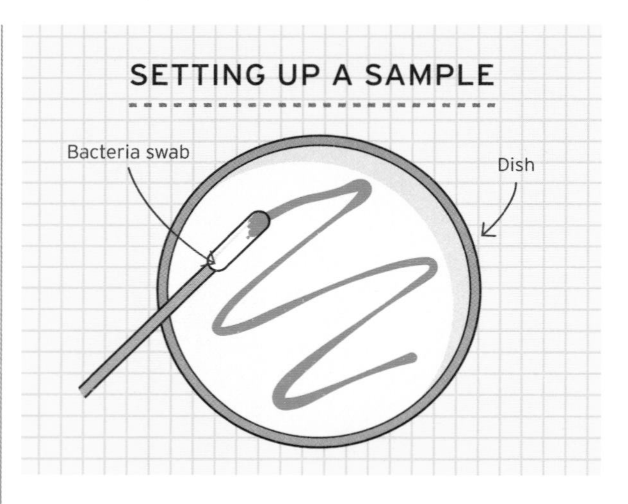


## WHAT TO DO:

**1.** Boil the water for at least a few minutes, to kill any bacteria which might already be in the water—they can't survive in high temperatures.

**2.** Mix the gelatin, broth (or yeast extract), and sugar in the pitcher, and then add the water (ask an adult to help), and stir until it's dissolved. Pour some of this liquid into each dish, cover them with a lid or plastic wrap, then refrigerate overnight.

**3.** Once the gelatin is set, it's time to introduce some bacteria. Use a separate cotton swab and dish for each test. You could swab your mouth, your fingers, or anywhere that people touch—such as a phone, handle, or coins. Rub the swab on the item to pick up a sample, then swipe it gently across the surface of the gelatin to deposit the bacteria. You won't be able to see them, but they're there!

**4.** Cover the dishes with plastic wrap, and leave them somewhere warm for a few days.

## DON'T EAT IT!

While food products were used to grow the bacterial colonies, they are not safe to eat! There's no way to know which type of bacteria will grow, and some can make you sick if eaten. Label your experiment, wash your hands well whenever you've touched the samples, and dispose of your bacteria afterward to make sure nobody mistakes it for a snack!

## WHAT HAPPENS?

You should be able to see colonies of bacteria growing on the surface of the gelatin. When they reproduce, every bacterium divides in two, and eventually a colony is formed, with enough bacteria that you should be able to see it unaided. Even though it looks like one mass, it's actually many separate organisms. *E. Coli* bacteria divide every 20 minutes, so from one bacterium, in seven hours you'd have over two million!

## IDENTIFYING BACTERIA

This method is used by scientists to identify types of bacteria. Rather than using a microscope, they can culture the bacteria, growing millions, and identify it by the visible characteristics of the colony, called the morphology of the bacterium.

Cold, dry conditions and high levels of salt all inhibit bacterial growth—and these are two methods used to preserve food. In this experiment, low-sodium beef stock, water, and warmth encourage growth.

# LEARN ABOUT: BREAD AND YEAST

## Can you answer these yeasty questions?

### POP QUIZ: BREAD AND YEAST

**1.** Which of the following is used as a food source for yeast in food manufacturing?
a) Boiled potatoes
b) Sugar
c) Malted barley
d) All of the above

**2.** What's the meaning of the Latin name of the yeast species *Saccharomyces cerevisiae*?
a) "Baker's helper"
b) "Sugar and spice"
c) "Sugar-eating fungus"
d) "Sweet addition"

**3.** In which ancient civilization was the first recorded use of yeast to bake bread?
a) Ancient Egypt
b) Ancient Babylon
c) Ancient Rome
d) Ancient Greece

**4.** Which of the following are not made with yeast?
a) Donuts
b) King cake
c) Tortillas
d) Cinnamon buns

**5.** The process by which yeast converts sugar into carbon dioxide is called:
a) Fermentation
b) Carbonation
c) Saccharization
d) Omnomnomation

**6.** In which of these conditions can yeast cells not survive?
a) Too much salt
b) Too much sugar
c) Too much alcohol
d) All of the above

**7.** Which of these is not something yeast is used in?
a) Making non-alcoholic drinks
b) Baking soda bread
c) Treating diarrhea
d) Fish tanks and aquariums

**8.** If you use dried yeast, how do you "activate" it before use?
a) Shock it with electricity
b) Soak it in warm water
c) Read it the mission briefing
d) Add plenty of salt

**9.** Nutritional yeast, which has been deactivated (killed), is sometimes added to food to:
a) Add a cheesy or nutty flavor
b) Produce bubbles of carbon dioxide to make bread rise
c) Add color
d) Increase levels of salt, fat, and sugar in the food

**10.** Yeast is classified as:
a) A bacterium
b) A mammal
c) A virus
d) A fungus

# **LEARN ABOUT:** BACTERIA AND MOLD

**There are numerous varieties of bacteria and mold, many of them present in the environment around us. But how much do you know about them? Test your knowledge with these challenges.**

## POP QUIZ: BACTERIA AND MOLD

**1.** Out of the following, circle any conditions which would make bacteria grow more quickly, and cross out any that would inhibit the growth of bacteria:

| | | |
|---|---|---|
| Inside a hot oven | Moist surroundings | Low salt levels |
| A warm environment | Dry surroundings | Sugars, starches, and proteins present |
| In a freezer | High salt levels | No sugars, starches, or proteins present |

**2.** Which of these properties are true of bacteria, and which are true of mold? (Careful: some of them apply to both... or neither!)

| | BACTERIA | MOLD |
|---|---|---|
| Single-cell organism | | |
| Forms long filaments | | |
| Reproduces by dividing | | |
| Reproduces by releasing spores | | |
| Considered a microbe | | |
| Comes in many different shapes and forms | | |
| Causes food to spoil | | |
| Typically wears a hat | | |
| Can be used in making cheese | | |
| Too small to be visible unless allowed to grow | | |
| Cannot grow when the temperature is too low | | |

# EXPERIMENT: MAKING YOGURT

**Not all bacteria are harmful. In fact, some types of bacteria are used in making foods, such as yogurt. It's so easy, you can even do it at home!**

You might think that finding the right type of bacteria would be tricky, but it turns out that one of the basic ingredients of yogurt... is yogurt! It's the best place to find yogurt-making bacteria, but it needs to be live yogurt, which means it's not been heat treated (pasteurized), which would kill the bacteria.

### YOU WILL NEED:

- 2 liters (4 pt.) milk (skim, 1%, and 2% are all fine)
- 100 g (7 tbsp.) powdered milk (optional, if you want thicker yogurt)
- 100 ml (3½ fl. oz.) plain, live yogurt
- Saucepan
- Thermometer
- Mixing bowl
- Hand whisk
- Thermos flask or plastic wrap and clean towel
- Sterile airtight container (e.g., a mason jar or coffee jar)

**ADULT SUPERVISION REQUIRED**

### PASTEURIZATION

Heating milk will kill any bacteria in it—including the harmful ones—as bacteria can't survive at high temperatures. The process is known as pasteurization, named for its inventor, the scientist Louis Pasteur. In this experiment, the milk must be cooled before adding the yogurt, otherwise the heat will kill the helpful bacteria as well!

**WHAT TO DO:**

**1.** Heat the milk in the saucepan to around 80°C (187°F). (Ask an adult to help.) If you don't have a thermometer, take the milk off the heat when it's steaming and bubbling around the edges.

**2.** Pour the hot milk into the mixing bowl, and leave it to cool down to about 46°C (114°F).

**3.** Whisk in the live yogurt. (If you'd like a thicker result, also whisk in the powdered milk at this point.)

**4.** Now your yogurt needs to be left somewhere warm. You could pour it into jars and put it in an oven with just the pilot light on, or cover the top of the bowl with plastic wrap, place a towel on top, and leave it someplace warm in the house. Alternatively, you could swill some hot water around in the thermos to warm it, tip it out and put the yogurt in there overnight.

**5.** After 6–8 hours (or overnight), it should have thickened. Pour the yogurt into a sterile, airtight jar and refrigerate it.

## WHAT HAPPENS?

*Streptococcus thermophilus* and *Lactobacillus bulgaricus* are present in the yogurt you added. These harmless bacteria feed on the lactose sugars in the milk and break it down by fermentation, producing lactic acid.

As the experiment on pages 54-55 showed, yeast likes to be warm when it ferments things, and so does bacteria, which is why still-warm milk is used. In fact, the name "*thermophilus*" (from Greek: *thermos* means "hot"; *philos* means "loving") roughly translates to "likes warmth"!

The lactic acid gives yogurt its tangy taste, and also reacts with the proteins in the milk, making it thicken into yogurt.

# **DISCOVER:** FOOD DECAY AND FOOD POISONING

**While fungi, bacteria, and mold have many uses in producing food, they're also responsible for many types of food decay. Take care when storing food, as over time the decay can lead to the food losing nutritional value and tasting different, and it can contain toxins and microorganisms that might cause illness, injury, or even death if eaten.**

## FOOD SPOILAGE

Food can spoil for many reasons. Exposure to light, heat, and air can all change the color and taste of the food. Enzymes—proteins that cause chemical reactions in the food to happen more quickly—can continue to act if the food is left too long. For example, the enzymes that make a banana turn from green to yellow as it ripens will continue to work, and will eventually turn the banana black.

The most common type of food spoilage is due to microorganisms—bacteria, mold, and fungi. Fungal spores and bacteria can land on food if it's left open to the air, and the tiny quantities present on the food, given time and the right conditions, can start to multiply.

Some types of bacteria merely cause food to spoil. For example, when milk goes sour, it's bacteria fermenting the lactose in the milk into lactic acid, giving it a sour taste. But other types of bacteria are pathogens, which means they cause disease, and will make you sick if you eat them in large enough quantities.

It's hard to tell if food contains bacteria because they're too small to see. Raw meat, fish and shellfish, and dairy and eggs can provide an especially good breeding ground for bacteria, including *Salmonella* or *Escherichia coli* (*E. coli*).

Mold growing on one part of a food item usually indicates there are long filaments of mold growing into and through the food, especially if it's a soft, porous food like bread. Even if you cut off the part that's visibly moldy, there could still be mold present that you can't see.

## FOOD POISONING

Small quantities of microorganisms will be dealt with by your body's immune system, but if you eat food that's not been carefully handled, the bacteria can cause a serious infection.

Symptoms include nausea (feeling sick), diarrhea, vomiting, stomach cramps, a fever of 38°C (100°F) or above, tiredness, aches, and chills. The vomiting and diarrhea are designed to remove the offending bacteria as quickly as possible from your digestive system, and there's not much you can do but wait a few days until it passes.

Even though all this sounds awful, there's good news—spoilage and food poisoning are easy to prevent. Turn the page to find out how!

# DISCOVER: KEEPING YOUR FOOD SAFE

**While bacteria and mold can cause serious problems with food, there are plenty of simple things you can do to stop them.**

Since you can't see microbes on your food unless they've grown into a colony, you need to take precautions and not give them an opportunity to grow. Now that you know about the biology of bacteria, mold, and fungi, here are simple steps to prevent them from spreading.

**KEEP FOOD COOL** Bacteria and mold aren't killed by low temperatures, but it does stop them from multiplying. Keep fresh food cool in the refrigerator, or freeze it, to stop microbes from growing.

**COOK IT PROPERLY** Many microbes are killed by heat; it breaks down the proteins they're made from. When you cook food, you should make sure it reaches at least 73°C (165°F) all the way through to kill any bacteria present.

**DON'T BUY TOO MUCH** When you shop for food, plan when to use it and make sure none of it hangs around too long. Fresh food is labeled with "use by" dates to help you plan.

**BACTERIA** grow quickest when the temperature is between 4°C (40°F) and 60°C (140°F). This is the danger zone! To keep food safe, either keep it cool in the refrigerator or freezer, eat it while it's hot, or keep it hot in a slow cooker or chafing dish, if it's being served in a buffet.

**HANDLE IT GENTLY** Cracks and bruises on fruit and vegetables are often places bacteria and mold can grow more easily, and dented cans and broken packaging may have a reduced shelf life. Remember: your groceries are not footballs, so don't throw them around!

**KEEP IT SEALED!** Microbes like bacteria and mold mostly need oxygen to breathe, so if you wrap your food tightly, it'll stop them from getting in, and also stop any bacteria already in there from getting access to oxygen. You can use cans and jars with lids, sealable plastic tubs, and waxed cloth wraps to keep things airtight—or put a plate on top of a bowl.

**WASH YOUR HANDS** You might feel like you've been hearing this since you were a little kid, but washing your hands is so important in preventing food poisoning! Bacteria like *E. coli* and *Salmonella* naturally occur in the human gut, and people not washing their hands after using the bathroom is the main way they make it into your food—either directly, or by transfer from things you've touched. Washing your hands before preparing or eating food, or after touching anything dirty, is essential to stop infections, especially when you're sick.

**DON'T TAKE CHANCES** If food has been left out in temperatures over 4°C (40°F) for more than two hours, you should throw it away. Even if you're planning to cook it later, microorganisms on food produce toxic chemicals, and even if the bacteria or mold is destroyed by heat, the toxins remain and can cause indigestion or vomiting. If you aren't sure, smell the food before eating. If it smells bad, that's a sure sign it shouldn't be eaten.

# EXPERIMENT: FOOD TESTS

**The main substances present in food are proteins, sugars, starches, and fats. Various tests can be carried out to identify which of these are present in a food sample, and some of them you can do at home! Check out pages 94–95 to find out more about each of these nutrients.**

## PROTEINS

Proteins are made up of smaller pieces called amino acids and are essential for many biological processes, from cell repair to replication; enzymes, DNA molecules, and many structural parts of cells are all proteins.
**PRESENT IN:** meat, fish, dairy, eggs, seeds, nuts, beans, and other legumes

Scientists can use Biuret reagent to test for protein. It's a mixture of sodium hydroxide and copper sulfate, which is normally blue. However, when it reacts with protein, it changes color to pink-purple/mauve.

COPPER SULFATE CRYSTALS

## SUGARS

Sugars are used by your body for respiration (see pages 32–33), and of the many types, glucose is the most common. Fructose is found in fruits, lactose in milk, and sucrose (made of glucose and fructose) occurs naturally in plants such as sugarcane.
**PRESENT IN:** fruit, candy, processed foods, sauces, juices, and drinks

The scientific test for detecting simple sugars also uses Benedict's reagent, the mixture of copper sulfate, sodium citrate, and sodium carbonate. If added to sugar, it changes from blue to green, yellow, or orange-red, depending on the amount of sugar present.

## STARCH

Starch molecules are long chains of glucose molecules joined together. Unlike glucose, they don't taste sweet, but are a good way to store energy.
**PRESENT IN:** bread, rice, pasta, potatoes, cereal, and oats

## TESTING FOR STARCH

If you want to test whether a food contains starch, the easiest way is to use iodine. You can buy it in bottles from a pharmacist—it's used as an antiseptic. While iodine solution is usually orange in color, it will turn dark purple in the presence of starch. Use a drop to test small samples of food (but don't eat them afterward!).

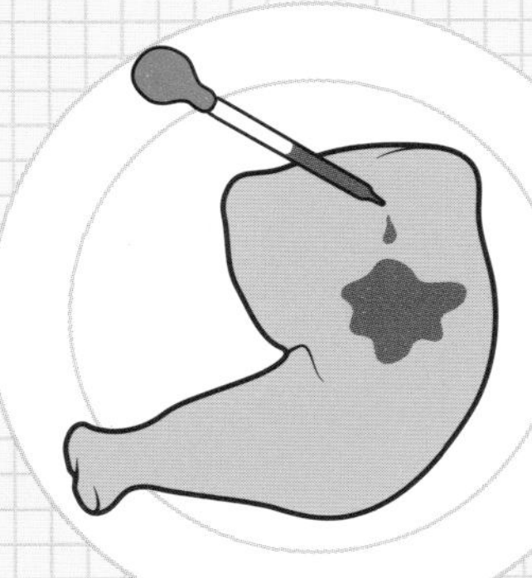

## FATS

Also called lipids, these contain a lot of energy and can be used by the body for energy storage, and as part of cell membranes. Lipids dissolve in alcohol but not in water. This is why oil and water don't mix: because oil is made up of fat molecules.

**PRESENT IN:** oils, dairy products, eggs, and nuts

**TO TEST** whether a solid food contains lipids, use a piece of paper. If you rub the paper on the surface of the food and a translucent patch forms on the paper, that means it contains lipids.

# DISCOVER: WHAT'S IN AN EGG?

**Eggs are used in cooking, but they're also biologically fascinating. Egg-laying animals are called oviparous, and include birds, reptiles, amphibians, and most fish. An egg is made from a single cell, and will grow into a baby animal if it's fertilized.**

## WHY EGGS?

Oviparous creatures reproduce using eggs, which are fertilized before or after being laid. The baby develops inside the egg but outside of its mother's body—unlike mammals, which grow in a womb.

An unfertilized egg won't grow into a baby. In humans, unfertilized eggs are removed from the body along with menstrual discharge each month, and chicken eggs are sold unfertilized, which means they don't contain a growing chick (and the eggs are checked for anything growing inside by shining light through them).

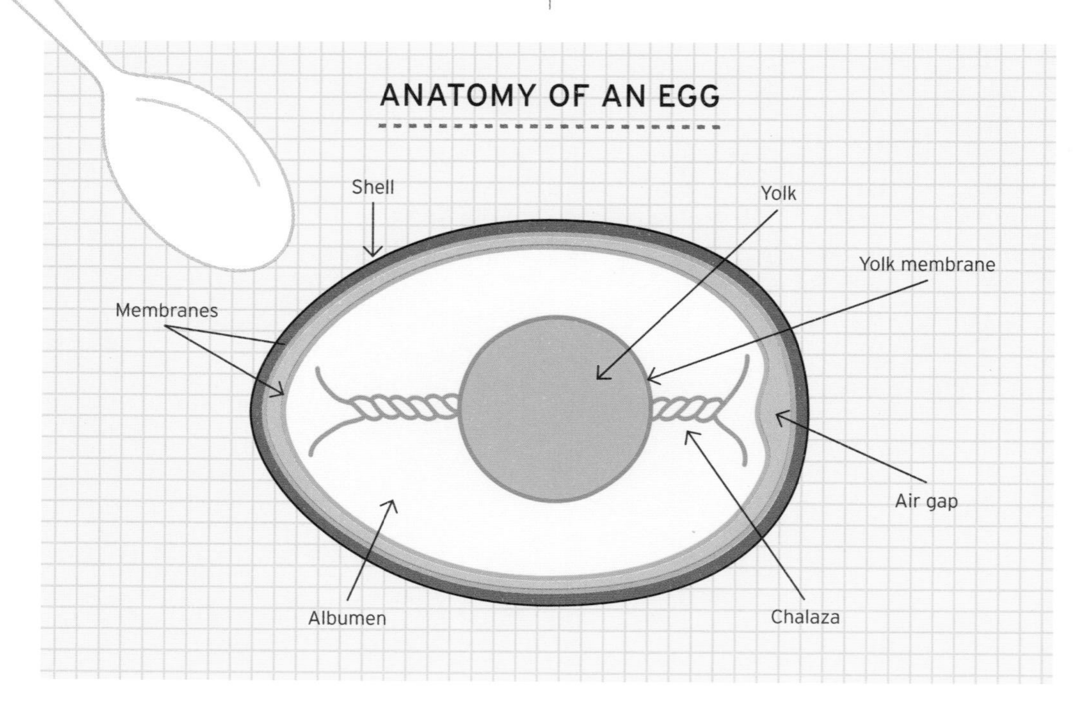

## WHAT'S IN A CHICKEN EGG?

The shell of an egg is made from calcium carbonate, inside which are membranes made from keratin (the same protein your hair is made from!). These keep out bacteria and dirt, and even though they're solid, they're semipermeable, which means they allow air and moisture through. There are around 17,000 pores on the surface of an egg!

Inside is the albumen, called egg white, made from water with a little protein; inside that is the egg yolk, within another membrane. The yolk is high in fats, protein, vitamins, and minerals, and serves as a food source for a developing chick.

### OTHER KINDS OF EGGS

As well as hen's eggs, people also eat other eggs, including duck, turkey, and goose (thicker shell, larger yolk, and richer taste), quail (tiny and light-tasting), and ostrich, which are huge and weigh around 1.3 kg (3 lb.), making them the largest single cell in biology! Emu eggs are black with green speckles, and their contents are so thick they don't run out if you cut the egg open. Fish eggs, called roe, are also eaten as food.

The egg also contains a blastodisc, a clump of cells that would develop into the chick if fertilized, which is visible as a dot on the surface of the yolk. There's also an air gap at one end of the egg, formed when the contents of the egg cool and contract after the egg is laid.

**OVER TIME**, the air gap in an egg can grow, and this means if you place an egg in plain water, it'll sink to the bottom if it's fresh, and float to the surface if it isn't. Some people use this as a test to see if an egg has gone bad. While it can give you a clue as to how fresh the egg is, the best way to check is to crack it open and see if it smells bad!

Eggs also contain a piece of stringy white spiral tissue called the chalaza, which holds the yolk in place. It gets twisted around when the yolk of the egg rotates, and it connects to the membrane surrounding the yolk like a bag. It's safe to eat, but some chefs take them out to keep their sauces smooth.

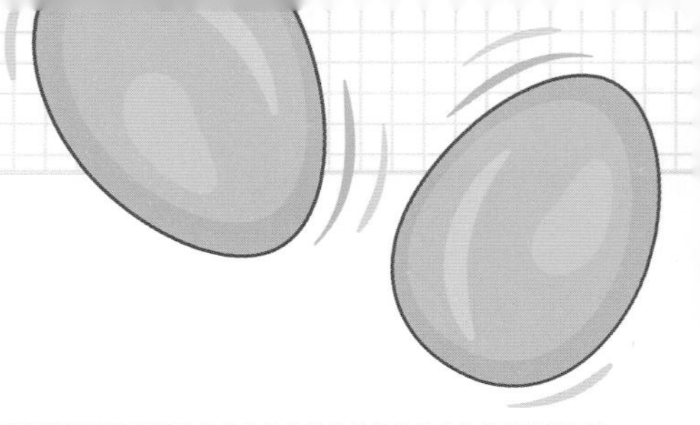

# EXPERIMENT: MAKING A RUBBERY EGG

**Vinegar contains a weak acid called acetic acid, and since acids react with carbonates like the calcium carbonate found in egg shells, these two kitchen staples can be used to perform a fun and easy science experiment: making an egg into a rubber ball!**

## YOU WILL NEED:

- **1 egg—it doesn't have to be cooked, but if you want to be sure it's not going to cause a mess, you can boil it first**
- **1 small glass**
- **White vinegar (enough to cover the egg in the glass, twice)**

## WHAT TO DO:

**1.** Put your egg in the glass and cover it with vinegar. Label the glass so nobody drinks it—they should notice the smell of vinegar, but better safe than sorry—and place it somewhere for 24 hours.

**2.** Foam should have appeared around the egg. Handling it carefully, rinse the egg off with water, and replace it in the glass covered with new vinegar.

**3.** Wait for six days, then rinse off the egg.

## WHAT HAPPENS?

The egg feels different (compare it to a fresh egg). The shell has gone, and the egg feels rubbery.

Investigate by:

- Shaking the egg
- Squeezing the egg (gently!)
- Dropping it from a height of a few inches, into the bottom of the sink
- Once you're confident, dropping it from higher up
- Rolling it along a table

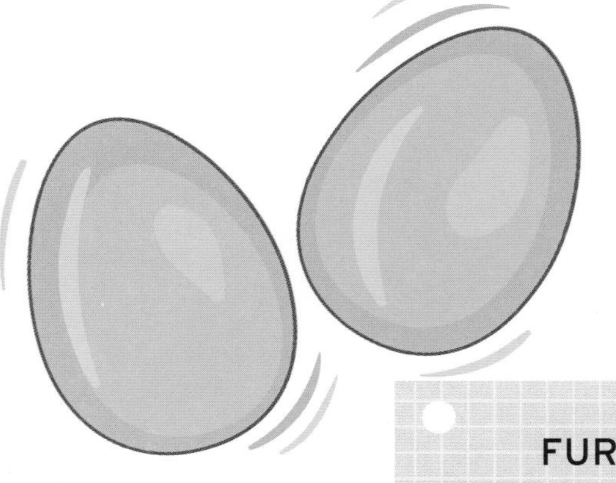

### WHY IS IT RUBBERY?

The shell of the egg is made from calcium carbonate, which is dissolved by the acetic acid present in the vinegar. The initial foaming reaction is the same kind of reaction you get if you mix vinegar and baking soda—an acid mixing with an alkaline (carbonate)—and the bubbles will be the carbon dioxide gas produced by the reaction.

The shell will be fully dissolved by the vinegar, and the semipermeable membrane underneath will be exposed. Since the concentration of vinegar inside the egg is lower than outside, osmosis will occur and vinegar will pass through the membrane into the egg, turning it into a pickled egg! This makes the membrane of the egg tougher, so it can be bounced and rolled like a ball (but it can still break, if you're too rough with it!).

## FURTHER INVESTIGATIONS

- What happens if you leave the egg for a longer or shorter time? Does soaking it longer mean it will bounce higher? Measure bounce height with a ruler, making sure you throw the egg downward with the same force and from the same height. (Remember to only change one variable in your experiment!)

- Does the weight or size of the egg change when you soak it in vinegar? Make a prediction, and measure it before and after soaking to find out if you were right!

# CHAPTER 3
# **YOU**

DISCOVER...

LEARN...

EXPERIMENT...

# **DISCOVER:** YOUR DIGESTIVE SYSTEM

**While all the cells of your body use the energy and nutrients you get from eating food, the main part of you that interacts with food is your digestive system. This consists of the gastrointestinal tract—the route food takes through your body—and a few other organs, which create the enzymes and hormones used for digesting food.**

### MOUTH

Your mouth is the first port of call for the food you eat. It contains teeth to break up food into smaller pieces, and glands that produce saliva to lubricate food in your mouth and start the digestion process. There's also a tongue to manipulate food around and shape it into a ball, called a bolus, which you can swallow using the muscles of the tongue and pharynx (the back part of the mouth).

### ESOPHAGUS (GULLET)

This is the pipe that runs from the back of the mouth down to the stomach. Using a series of rings of muscle which contract one at a time, the food is pushed down toward the stomach using a process called peristalsis.

## STOMACH

This is a J-shaped organ connected at the top to the esophagus. Here, food is churned by more peristaltic muscle contractions, mixed with strong acid, and broken down further by enzymes released by glands in the walls of the stomach. The stomach can stretch to hold about 1 liter of food, and food stays here for an hour or two, after which it becomes a semiliquid called chyme.

### THE LIVER, GALLBLADDER, AND PANCREAS

The liver, gallbladder, and pancreas are all part of the digestive system, but food doesn't pass directly through them. They produce the bile and enzymes used in digestion; the liver, a large triangular organ on one side of the body, also breaks down toxins found in the food, and converts various types of nutrients to other types, as needed to regulate levels in your body.

## SMALL INTESTINE

The chyme is released from the stomach by opening the pyloric sphincter, a ring of muscle that can contract to close, and relax to open. It flows into the duodenum, the first part of the small intestine, where bile from the gallbladder (a small pear-shaped pouch) is added to neutralize the stomach acid. To break down fats and proteins, more enzymes from the pancreas (a flat organ tucked away at the back, near the bottom of the stomach) are added, some of which won't work in the acidic conditions of the stomach. Nutrients from the food start to absorb through the walls of the intestine into the bloodstream.

The mixture then passes into a second part of the small intestine called the jejunum, and then to the ileum (the final part of the small intestine), where more enzymes are added to break down starches into sugars. The muscles of the walls of the intestine contract to push the food along, and to mix it with the enzymes. The small intestine winds back and forth across the abdomen, and the internal walls are covered in tiny lumps called pili, which give it a large surface area for absorbing as much nutrition as possible.

### LARGE INTESTINE (COLON)

Once the food passes through the cecum (which joins the small intestine to the large intestine), through the ileocecal valve which controls the flow, most of the nutrition has been extracted. It's now around 6–8 hours since you ate the food, but it'll take another 15–20 hours to pass through the large intestine. It still contains indigestible fiber which gives it bulk, as well as water. In the colon, this water is absorbed out and the food is gradually reduced to a solid substance, which passes into the rectum and is pushed out as feces, through one final sphincter called the anus.

### OTHER PARTS

The appendix is attached to the intestines, where the small intestine meets the large intestine. It's a finger-shaped pouch that hangs off the bottom corner of the cecum, and its purpose is still something of a mystery—if your appendix gets infected, it's incredibly painful, but it can be surgically removed and this seems to have very little effect.

It used to be thought the appendix had previously been used by humans to digest cellulose, but that it had gradually reduced in size and become useless as humans adapted to eating a less plant-heavy diet. This idea has now been replaced by other theories, suggesting it has a role in regulating bacteria in the body.

The diaphragm is a sheet of muscle that divides the thorax (the upper part of your internal body cavity) from the abdomen, where all the digestive organs are. It's very important in controlling the flow of food through the digestive tract, and contractions of the diaphragm are used to push digesting food around. If you ever get the hiccups, it's because the muscles in your diaphragm are contracting involuntarily, in response to irritation of the nerves.

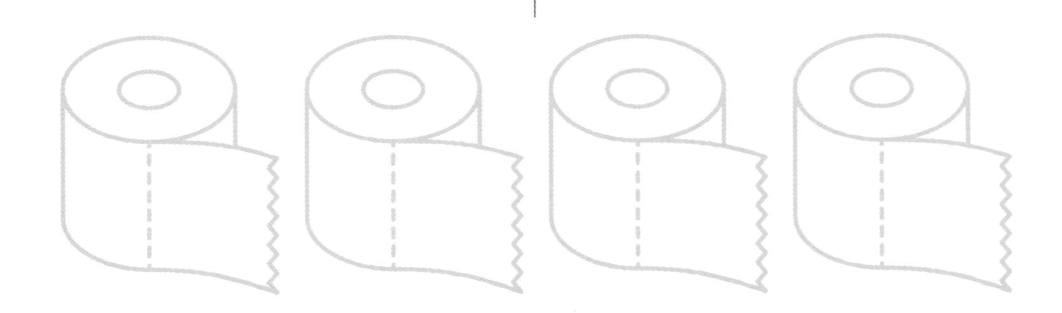

# **LEARN ABOUT:** PARTS OF THE DIGESTIVE SYSTEM

Can you identify where all the parts of the digestive system are on this diagram, using the descriptions given on pages 84–86?

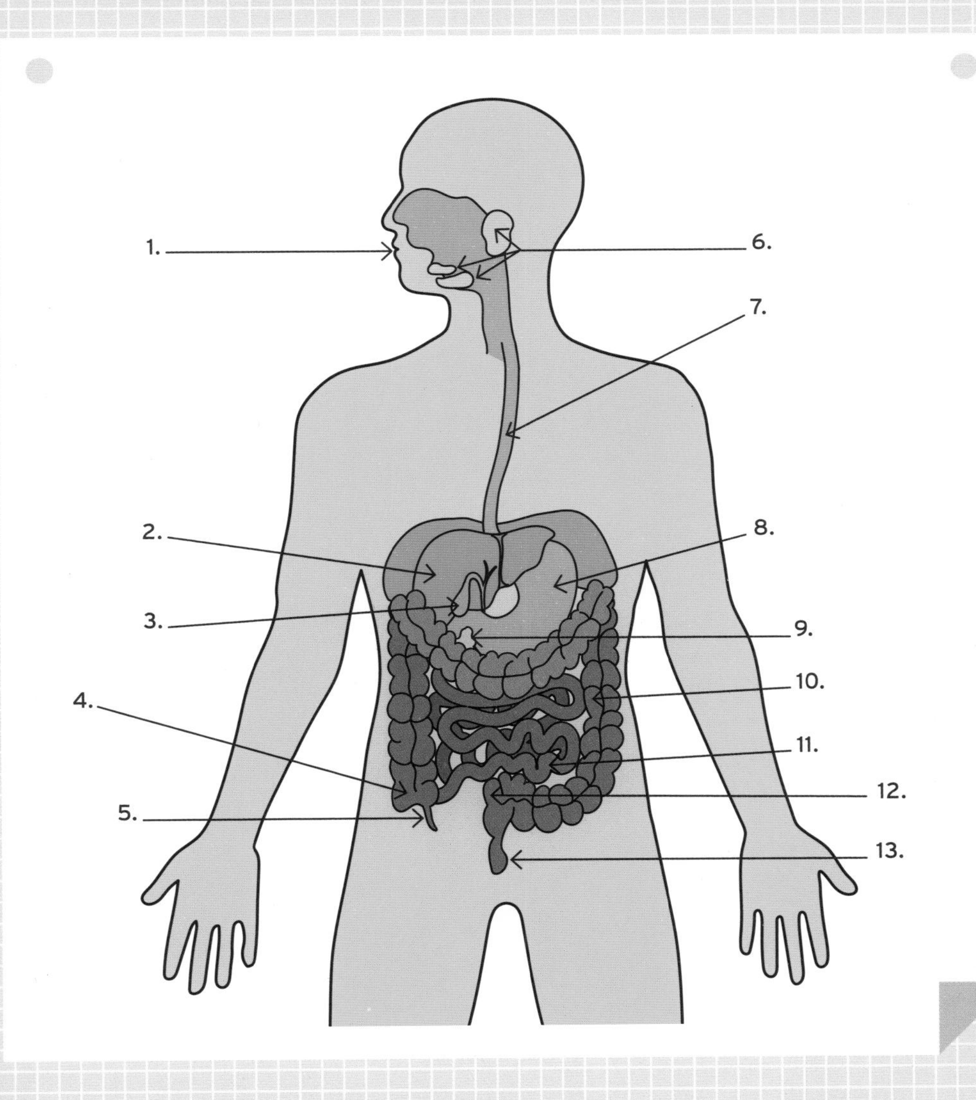

# EXPERIMENT: ENZYMES IN YOUR SALIVA

**Food is digested as it passes through your body, but the process of digestion actually starts as soon as you put the food in your mouth. The saliva your glands secrete doesn't just act as a lubricant to mush up your food, but it contains an enzyme called amylase, which speeds up the process of breaking down starches into sugars.**

This simple experiment tests a little saliva to see what effect it has on starch.

### YOU WILL NEED:

- White bread
- Your mouth
- A sealable container

### WHAT TO DO:

**1.** Take a piece of bread and put it in your mouth. Notice what it tastes like; it should not taste very sweet.

**2.** Chew the piece of bread for 30 seconds—you should notice it starts to taste sweeter.

This is because the amylase enzymes in your saliva break down the starch in the bread into individual sugar molecules, which taste sweet.

**3.** Take a second piece of bread (you can swallow the first piece if you want) and chew it for about ten seconds. Spit the bread, along with a bunch of saliva, into the container, seal it and leave it somewhere at room temperature for about ten minutes.

**4.** Take out the bread and taste it. It should taste much sweeter because the amylase enzymes have had time to break down even more of the starch in the bread and turn it into sugars.

## A MORE SCIENTIFIC APPROACH

If you have access to iodine solution (perhaps from the experiment on page 77), you could use it to investigate this amylase reaction further.

- An unchewed piece of bread should test positive for starches, making the iodine solution turn from orange to dark purple.
- If you briefly chew a small piece of bread (not one with iodine on it) and place it on a sample dish, then add a few drops of iodine, it should react less strongly.
- If you chew another piece of bread and leave it in a sealed container for a while, you should find even less starch is present, and the iodine reaction will be much less strong.

Always make sure you test your samples on a plate—don't put any iodine in your mouth! (As well as tasting awful, it's much harder to see what color it's changed to if it's in your mouth.)

Amylase enzymes react differently under different conditions. Can you design an experiment to test what effect pH has on the action of amylase, using an acidic substance like vinegar, or an alkaline one like baking soda, and testing how quickly it breaks down the starch and turns the iodine back to orange? What about testing the effect of temperature: can you find the optimal temperature for amylase? What do you predict it might be? Don't forget to include a control sample to compare it against!

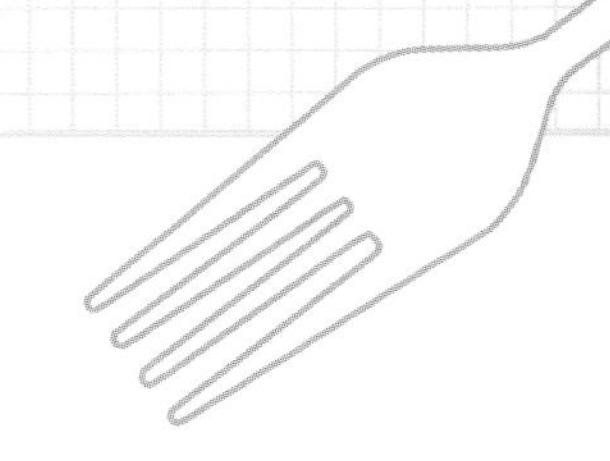

# **EXPERIMENT:** EXTRACTING YOUR OWN DNA

**DNA is found inside all of your cells and contains instructions that tell them how to function and reproduce. The information is in the form of a very long sequence of nucleotides—four different types (adenine, cytosine, thymine, and guanine) arranged along a long molecule, which is twisted into a tiny ball in the nucleus.**

In this experiment, you'll break down the walls of some of your cheek cells, and extract the DNA strands so they uncoil and can be seen.

### YOU WILL NEED:

- 2 clear glasses
- 20 g (4 tsp.) salt and 50 ml (3 tbsp.) water
- Liquid soap
- 100 ml (3½ fl. oz.) of strong alcohol-based disinfectant chilled in the freezer for a few hours
- Food coloring

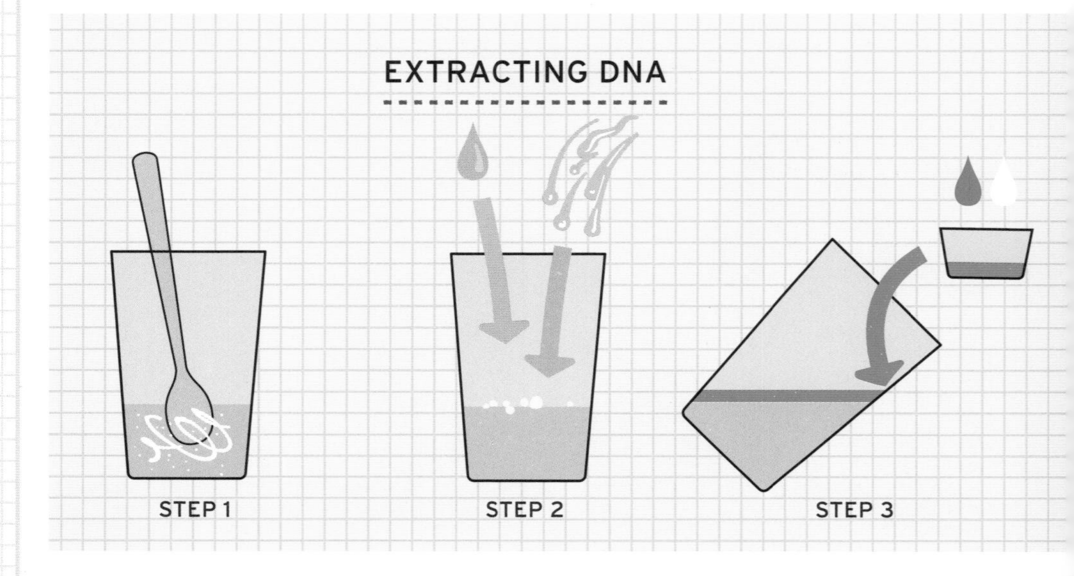

## WHAT TO DO:

**1.** Mix the salt and water in one of the glasses, stir until it's dissolved, and gargle it in your mouth for at least 30 seconds, swilling it around and using your teeth to scrape the walls of your cheek to loosen off some cells. (It's best to do this when your mouth is relatively clean—so, not right after eating.)

**2.** Spit the mixture back into the other glass, and add a drop of liquid soap. The soap will interact with the lipid molecules in the cheek cell walls, and break down the cell walls, releasing the contents of the cells into the water.

**3.** Add the food coloring to the chilled alcohol in a separate container. Tilt the glass containing the cells to 45 degrees, and carefully pour the chilled alcohol down the inside, so it forms a layer on top of the salt water.

**4.** Wait for 2–3 minutes. You should start to see clumps of white stringy mass forming in the alcohol layer—these are the uncoiled DNA molecules. DNA isn't soluble in alcohol, so they'll only be visible in this layer—they'll dissolve in the salt water. You can use a small skewer or toothpick to wind some of the molecules around and lift them out to look at. If you have a microscope, put them on a slide to look at more closely.

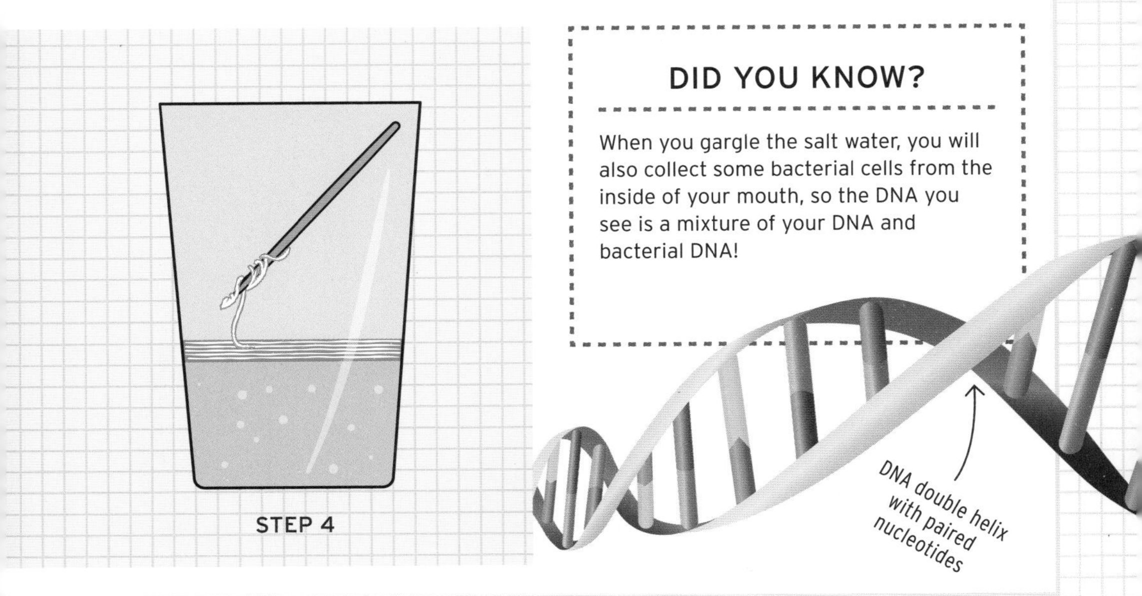


### DID YOU KNOW?

When you gargle the salt water, you will also collect some bacterial cells from the inside of your mouth, so the DNA you see is a mixture of your DNA and bacterial DNA!

# EXPERIMENT: BUILD DNA FROM CANDY

**Even though DNA is present in all living things, from tiny bacteria up to giant creatures like elephants, it always has the same structure. Build a model of a DNA molecule to see what it looks like, and use different colored candy to represent the different nucleotides.**

## THE STRUCTURE OF DNA

DNA molecules form a double helix structure: two long backbones made of sugar (deoxyribose) and phosphates are wound around each other, and nucleotides form bridges between the two sides. If you were to untwist the DNA molecule, you'd end up with one long strip that looks like a ladder, with each rung made up of two nucleotides: either an adenine–thymine pair, or a cytosine–guanine pair. The nucleotides always pair up with the same partner, and the order of A-T-C-G down one side of the DNA determines the order down the other side.

The sequence of nucleotides along the DNA molecules is used to encode information, which provides instructions for your cells on how to grow and behave. Unless you have an identical twin, nobody in the world has exactly the same DNA as you, and all of it is stored inside the nucleus of every one of your cells.

When DNA is replicated, the two halves of the molecule "unzip," separating between each pair, all the way down, then new A, T, C, and G nucleotides come in to form a new copy of each half, by matching with their corresponding pair.

## YOU WILL NEED:

- **Long, flexible candy rope**
- **Toothpicks**
- **Soft candy (something that comes in 4 different colors, like gumdrops or jelly beans)**
- **4 cups to separate candies by color, labeled "A," "T," "C," and "G"**

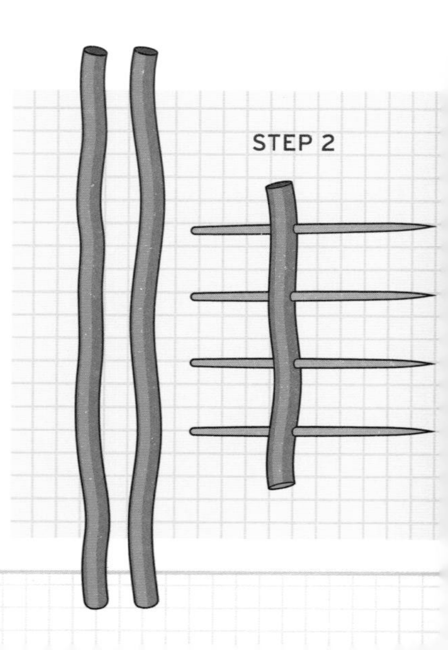

## WHAT TO DO:

**1.** Lay out two sugar-phosphate backbones (pieces of candy rope) down on the table in front of you. You're going to build a "twisted ladder" structure to represent the double helix.

**2.** Push toothpicks almost all the way through each of the sugar-phosphate backbones (candy rope), with regular spacing down each side. Lay the sugar-phosphate backbones (candy rope) down so that the tips of the two toothpicks touch in the middle at each point.

**3.** Now add one nucleotide (soft candy) to each toothpick, pushing it along to where the toothpick comes out of the sugar-phosphate backbone (candy rope).

**4.** Once you have completed one strip, make another one opposite. Don't forget to always pair the colors correctly: A with T and C with G!

**5.** Push the tips of each toothpick into the nucleotide (soft candy) opposite to connect the two halves together.

**6.** Once you've assembled your strip, pick up both ends and carefully twist it around into a double helix!

### REPLICATION MEANS MORE CANDY!

If you want to make your DNA replicate, pull it apart down the middle and then add in the correct colors of nucleotides (soft candy) to match with each one. Finish off by adding two more sugar-phosphate backbones (candy rope) and pushing toothpicks through to hold it in place.

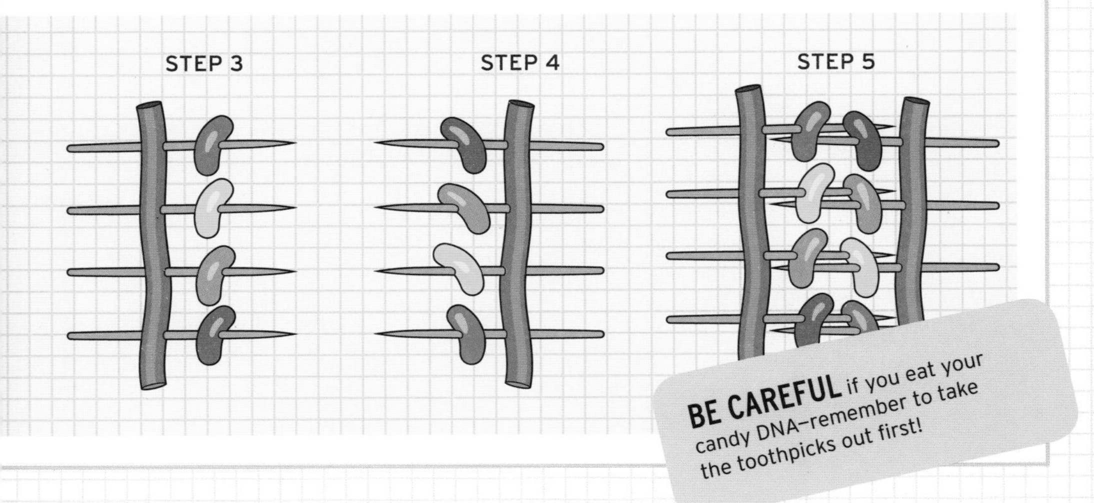


**BE CAREFUL** if you eat your candy DNA—remember to take the toothpicks out first!

# **DISCOVER:** FOOD AND DIET

**It's sometimes hard to decide what to eat—but what does your body actually need in your diet, and what does it use it for? The nutrition we get from food is categorized into several types, many of which you've already encountered on previous pages, and a good diet will contain the right amount of each.**

## TYPES OF NUTRIENTS

### Water

The most important thing in your diet is water: your body is 60% water, and it's used for dissolving and transporting substances and removing waste. It also helps keep you cool through sweating. It's found in liquids you drink, but is also extracted from solid food by digestion.

### Carbohydrates

Sugars are the building blocks of carbohydrates—a monosaccharide is a single sugar molecule with the formula $C_6H_{12}O_6$, with the form of sugar you get (like glucose, fructose, or galactose) depending on how the atoms are arranged. Two monosaccharides join to form a disaccharide, like sucrose or lactose. Many sugars joined together form a polysaccharide (a long chain of sugar molecules), such as starch or cellulose.

Carbohydrates are found in fruits and processed foods, grains (often crushed into flour for bread and other baked goods), vegetables, and dairy and milk products (in the form of lactose). Carbohydrates are the body's main source of energy.

### Proteins

Proteins are made of amino acids, composed mainly of carbon, hydrogen, oxygen, and nitrogen. There are 21 amino acids in the human body, and different proteins are made from different sequences of amino acids.

Proteins are used for growth and repair, including in muscle tissue, hair, nails, and skin. Proteins you eat are broken down into amino acids and rearranged into new proteins. They can also be converted into carbohydrates by the liver. Protein-rich foods include meat, fish, dairy and eggs, seeds, nuts, beans, and other legumes.

### Fats

Fats consist of long chains, called fatty acids, joined to glycerol molecules (and sometimes other components, depending on the type of fat). They are hydrophobic, which means they don't mix with water. They're another source of energy for the body, and they can be used to store energy, make cell walls, and transport fat-soluble molecules.

### Fiber

Also called roughage, this is indigestible matter like cellulose from plants. It helps the food you eat move through the digestive system.

### Vitamins and minerals

These are the organic and inorganic substances which the body needs in small quantities in order to function properly, but can't synthesize for itself in sufficient quantities. They're used in many chemical processes and in building molecules.

**VITAMINS** are named using letters of the alphabet. Ascorbic acid, known as vitamin C, is involved in cell repair and is important for the functioning of the immune system. Humans need 13 vitamins in their diet: A, B1, B2, B3, B5, B6, B7, B9, B12, C, D, E, and K.

**MINERALS** needed by the human body include potassium, sodium, calcium, magnesium, iron, zinc, copper, iodine, selenium, and cobalt. Some are needed in larger quantities—to build bone, transport oxygen in the blood, and send messages through the nervous system—while others are only needed in trace amounts.

# **EXPERIMENT:** WHAT'S IN YOUR DIET?

**It's so easy to eat anything that's put in front of you, and lose track of what you're consuming. Keeping a food diary is a good way to get a true picture of what you're eating, and make sure you're getting enough of the right kinds of foods!**

Keep a list of all the foods you eat in one day. List the foods, roughly how much, and at what time of day you ate them.

In order to make it a fair experiment, you shouldn't change your eating habits for the day; you want to get a true picture of what you normally eat, so don't let the fact that you're recording it affect your habits. If you're worried you can't do this, you could ask a friend or family member to observe and record what you eat, without telling you which day they're doing it.

## HOW GOOD IS YOUR DIET?

Government dietary guidelines recommend your diet should have:

• A variety of different types of nutrient-rich food. You need to make sure your diet contains plenty of vitamins, minerals, and amino acids, and there are many types, which you need a balance of. Eating the same foods all the time can mean you don't get the full range (and it sounds pretty boring!).

• Sources of protein: this can be seafood, lean meat and poultry, eggs, legumes (beans and peas, including soy products), nuts, and seeds.

• Fruits, especially whole fruits. Drinking juice, even freshly squeezed, means you miss out on the fiber you'd get from the whole fruit.

• Vegetables from a range of categories: dark green (leafy greens and broccoli), red and orange (bell peppers and tomatoes), legumes, starchy vegetables (potatoes), and others.

• Grains, including whole grains. White bread flour is made by crushing the grains and throwing away the husks, which contain fiber and nutrients, so try to also eat whole wheat bread and pasta (which are made without throwing anything away) or other whole grains.

• Dairy, including milk, yogurt, cheese, and/or fortified soy beverages; this is a good source of calcium, and if possible should be low-fat or fat-free.

You can also find out how many calories are in the foods you've eaten (online databases of the number of calories per gram in different foods are easy to find, or you can look at the packaging)—you should be eating around 2,000–2,500 calories per day.

It's also recommended that less than 10% of your calories should come from sugars, and less than 10% from saturated fats. The proportion of these in your foods should also be marked on the label, or can be looked up online. You should also aim to have less than 2,200 mg per day of sodium—that's equivalent to about one teaspoon of salt.

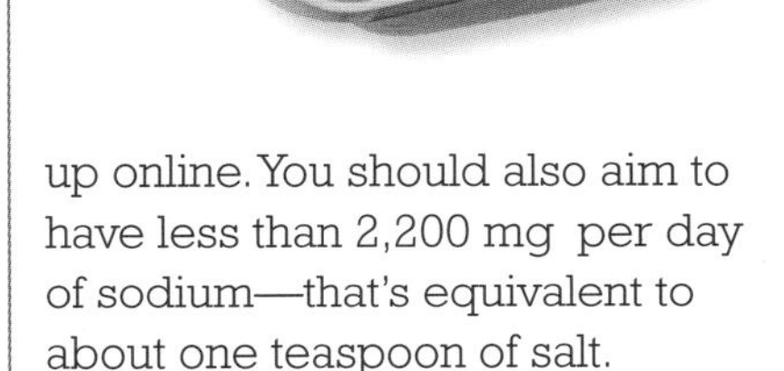

# **DISCOVER:** ENERGY IN FOOD

**A Calorie is a unit used to measure energy in food, and the number of calories in something determines how much energy you will get if you eat it. Calories can also be used to measure how much energy is used when you do a particular activity, so you can match up how much energy you consume with how much you burn.**

## WHAT IS A CALORIE?

One calorie is defined as the amount of energy needed to raise the temperature of one gram of water by 1°C (1.8°F). This means that one way to determine how much energy is in something is literally to set fire to it and use it as fuel to heat water. Carefully monitoring the number of degrees the temperature increases by can tell you the total number of calories of energy found in the food.

Another unit, the Calorie (with a capital "C") is defined as 1,000 calories. Calorie counts on food packaging are generally given in Calories, and most people use "calories," or "cal," to mean Calories, which is what you'll continue to see in this book.

Scientists typically measure energy in joules, where 1 joule is 1/4184 of a Calorie (and 4184 joules = 1 Calorie = 1000 calories).

## ENERGY IN FOODS

## HIGH-ENERGY FOOD TYPES

Energy in foods is mostly stored in the form of sugars and starches. Sugars are used in respiration to give your muscles energy to move, your neurons energy to send signals around the body, and all your cells the energy for chemical reactions, growth, and repair. Carbohydrates store around 4 cal/g (calories per gram).

Starch takes longer to release its energy; it needs to be broken down into sugars, and can't be immediately converted to energy like simple sugars can. This means it's a better way to get energy, and starchy foods often contain higher levels of fiber and other nutrients.

Foods with high levels of sugar will contain a lot more energy than the same amount of protein-rich or starchy foods. It's easy to consume a lot more calories than you realize—a large sugary drink can contain as many calories as the whole of the rest of your meal.

Fats can also be used by the body for energy and are much denser, storing around 9 cal/g. They are broken down into fatty acids and glycerol by a process called lipolysis. The fatty acids can be broken down directly to get energy, or converted into glucose. The body can also store energy in the form of fat molecules, which can then be broken down later, if needed.

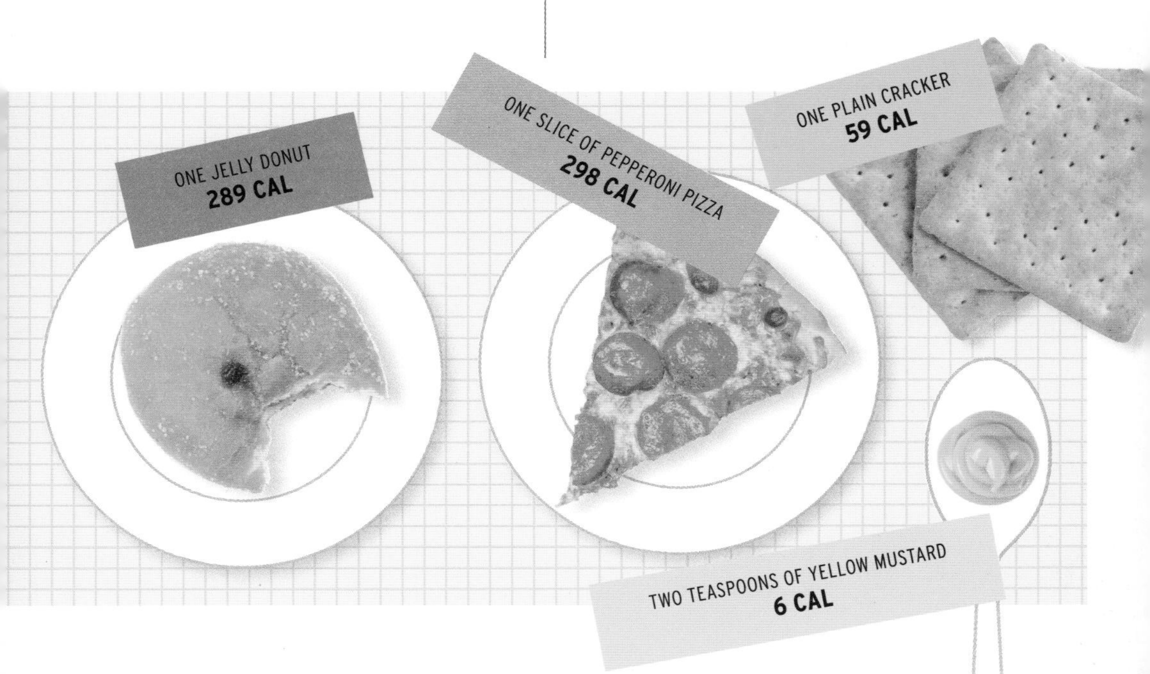

# DISCOVER: DIET DISEASES

**Having a balanced diet helps make sure you have the right amount of energy, nutrients, and minerals needed to maintain a healthy body. If your diet doesn't contain enough of the essentials, you might feel unwell or weak, and diseases might even develop. Here are some of the main deficiencies, along with their common symptoms.**

## PROTEIN DEFICIENCY

Mild symptoms can include tiredness and irritability. In extreme cases, this can extend to muscle wasting, hair and skin losing color, scaly skin, and water retention (swollen stomach). Protein-rich foods include milk, meat, and legumes (such as beans and peas).

## CARBOHYDRATE DEFICIENCY

The body can convert proteins and fats into energy if needed, but a lack of carbohydrate intake long-term can lead to ketosis, in which your breath smells weirdly sweet. Carbohydrates are found in starchy and sugary foods, but they're best consumed in foods that also contain fiber, such as fruit, vegetables, legumes, and whole grains.

## VITAMIN DEFICIENCY

Eating a range of different types of food will help you consume all the vitamins you need. Every type of vitamin is associated with some deficiency disease, with some more serious than others.

- **VITAMIN A:** Found in vegetables like carrots, sweet potatoes, and spinach; deficiency can lead to problems with night vision and, in more serious cases, blindness.
- **VITAMIN C:** Found in fruits and vegetables, like berries, bell peppers, and citrus fruits; deficiency is rare, but causes scurvy (symptoms include bleeding gums and bone pain). Vitamin C also helps fight disease, so a lack of it can lead to a greater incidence of getting sick.

• **VITAMIN D:** Found in fatty fish and dairy, and also created by exposure to sunlight; deficiency causes rickets, a disease of the bones which makes them soft and more prone to break.

## FORTIFIED FOOD

Some foods are fortified with vitamins and minerals, and some people also take vitamin supplements to make sure they're getting enough. But it is possible to have too much of a vitamin, and this is called hypervitaminosis. As long as you're getting a balanced diet (and unless you've been advised to do so by your doctor), you probably don't need to take supplements!

### MINERAL DEFICIENCY

Some of the most common nutritional deficiencies are due to a lack of essential minerals in the diet.

• **IRON:** Used in transporting oxygen around the body. Iron deficiency leads to anemia, which can cause fatigue, weakness, apathy, and decreased resistance to cold. Iron-rich foods include offal, dark green vegetables like spinach, and legumes.

• **IODINE:** Lack of iodine can cause goiters—swellings in the neck, caused by an enlarged thyroid gland. Iodine is found in seaweed, seafood, and dairy.

• **ZINC:** Zinc deficiency can lead to a range of problems, including acne and skin conditions, diarrhea, mouth ulcers, stunted growth, and decreased resistance to infections. Zinc is found in oysters and seafood, beef, legumes, and nuts.

**REMEMBER:** Every liter of your blood contains half a gram of iron—that's about as much iron as a medium-size iron nail!

# LEARN ABOUT: PICKING A GOOD DIET

**Take a look at these food diaries, and decide whether each of them meets the standards for a good diet by matching them up with these statements about healthy diets.**

**1.** This daily diet contains at least five servings of a variety of fruit and vegetables (where a serving is roughly the amount that would fit in the palm of your hand).

**2.** This diet has meals based on high-fiber, starchy foods such as potatoes, bread, rice, and pasta. (Potato chips and fries don't count!)

**3.** This diet contains some dairy, or dairy alternatives.

**4.** This diet has some protein—from beans, fish, eggs, or meat, for example.

**5.** This diet has enough water—at least 6 to 8 glasses of fluids a day.

**6.** This diet contains only small amounts of foods and drinks that are high in fat, salt, and sugar.

**7.** This diet has a variety of different foods from the five main food groups (vegetables/legumes, fruit, grains, protein-rich foods, dairy).

**8.** This diet meets the daily recommended total number of calories (2,000–2,500).

DIET B

## FOOD JOURNAL: Saturday

BREAKFAST: One glass of orange juice (54 cal), muffin (690 cal)

MORNING SNACK: Snack bag of chips (181 cal)

LUNCH: Sandwich with two slices of whole grain bread, turkey, and cheese (640 cal); apple (72 cal)

AFTERNOON SNACK: Can of cola (140 cal); chocolate bar (456 cal)

DINNER: Chicken (230 cal); large baked potato (280 cal) with two tablespoons of butter (200 cal); salad (15 cal) with dressing (310 cal); bread roll (330 cal)

1 2 3 4
5 6 7 8

DIET C

## THURSDAY'S food diary

BREAKFAST: Smoothie containing banana, oats, mango, milk, and honey (156 cal)

MORNING SNACK: Two multigrain rice cakes topped with peanut butter and jelly (218 cal)

LUNCH: Chicken bagel with mayo, lettuce, and tomato (521 cal); sugar-free soft drink (0 cal)

Afternoon snack: Apple (72 cal)

DINNER: Serving of carrot and lentil soup, served with flatbread (238 cal); small root beer float (330 cal)

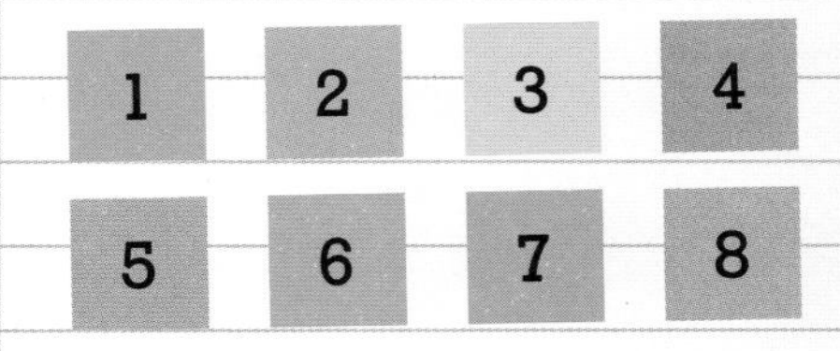


## MY EATING JOURNAL for Tuesday

DIET D

BREAKFAST: Egg sandwich (300 cal); serving of hash browns (140 cal); glass of water (0 cal)

MORNING SNACK: Apple slices (15 cal)

LUNCH: Five pieces of breaded chicken (630 cal); diet cola (0 cal)

AFTERNOON SNACK: Six cinnamon sugar donut sticks (280 cal)

DINNER: Quarter-pound burger with cheese (540 cal); small french fries (230 cal); glass of water (0 cal)

1 2 3 4 5 6 7 8

# DISCOVER: ENERGY AND EXERCISE

**As well as being a measure of how much energy is contained in food, the calorie also measures how much energy you need in order to perform activities.**

Energy is needed to make your muscles contract—skeletal muscles, which are used when you walk and move around, and also the muscles inside you, which make your digestive system work, cause your lungs to expand and contract when you breathe, and pump blood around your body. Depending on your size, your body will need different amounts of energy.

Find out how many calories these common activities burn up. (The figures on page 105 are all given for an average 12-year-old and are in calories per hour [cal/h].)

## NOT DOING MUCH AT ALL

Since your body is always performing various functions—breathing, digesting food, and sending chemical messages around the body through your nerves—it's always using a small amount of energy.

• Sitting quietly, not doing anything at all burns around 30 cal/h.

• If you're talking or watching something, you'll burn around 60 cal/h.

• Actively doing a moderate amount of work while sitting (moving things around, looking up and down), you'll get through more like 120 cal/h.

• Standing in place, you use about 90 cal/h, and even when you're asleep in bed you'll burn about 20 cal/h.

## OUT AND ABOUT

Simply walking at a moderate pace will crunch about 105 cal/h, but if you walk briskly and carry objects with you, that increases to around 290 cal/h. If the ground you're walking on is uneven, that uses around a third more energy than walking on flat ground. Cycling, even at a gentle pace of 5 mph, will burn 174 cal/h.

And if you find yourself going up stairs, think about running—you'll burn about 470 cal/h, more than the 380 cal/h you'd use to walk up (and much more than you'll burn if you take the elevator!).

## AT THE GYM

Many people go to the gym to burn calories, but you can also go there to build strength, stamina, and general fitness. There are so many activities to choose from, but here are a few of the popular ones:

- **RUNNING AT 6 MPH:** about 650 cal/h
- **JUMPING ROPE:** about 800 cal/h
- **VIGOROUS SWIMMING:** about 650 cal/h
- **AEROBICS:** about 400 cal/h
- **ROWING MACHINE:** about 350 cal/h

## HAVING FUN

You don't have to be a professional athlete to burn calories. More lighthearted and fun activities also take energy. You can burn around 274 cal/h playing volleyball, 384 cal/h in-line skating, 108 cal/h skateboarding, and even elegant ballet burns around 300 cal/h. And of course laughing burns calories: around 100 cal/h.

# **EXPERIMENT:** HOW MUCH SUGAR IS IN YOUR FOOD?

Sugar in food makes it taste sweet and appealing. That's why fast food, junk food, and processed food often contain high levels of sugar; it's an easy and cheap way to make food taste nice. But it's not much help if you're trying to eat a balanced diet! The levels of sugar in some foods can be surprisingly high, so it's important to check.

### YOU WILL NEED:

- A sample of food to analyze—the contents of your fridge, your kitchen cupboard, or, if you have kept one, a list of the food you've eaten in a day. Don't forget to include drinks too!
- Kitchen scale
- Pen and paper, and a calculator

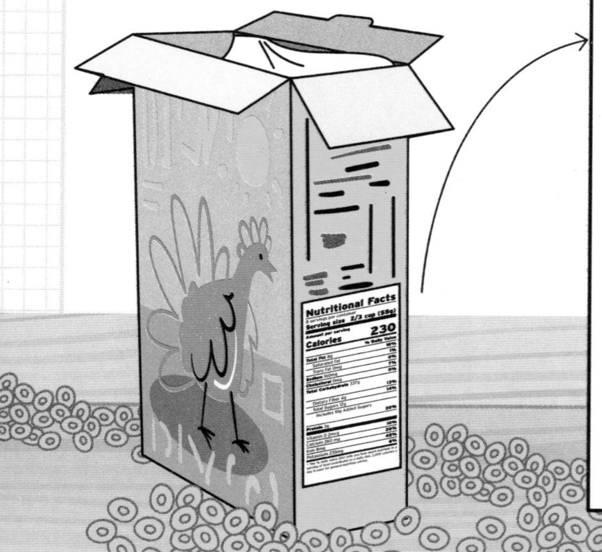

**Nutritional Facts**

8 servings per container

**Serving size 2/3 cup (55g)**

| **Amount per serving** | |
|---|---|
| **Calories** | **230** |
| | **% Daily Value** |
| **Total Fat** 8g | **10%** |
| Saturated Fat | **1%** |
| *Trans* Fat 0mg | **0%** |
| **Sodium** 160mg | **7%** |
| **Cholesterol** 0mg | **0%** |
| **Total Carbohydrate** 337g | **13%** |
| Dietary Fiber 4g | **14%** |
| Total Sugars 12g | |
| Includes 10g Added Sugars | **20%** |
| **Protein** 3g | |
| Vitamin D 2mcg | **10%** |
| Calcium 260mg | **20%** |
| Iron 8mg | **45%** |
| Potassium 235mg | **6%** |

* The % daily value (DV) tells you how much nutrient in a serving of food contributes to a daily diet. 2.000 calories a day is used for general nutrition advice.

**WHAT TO DO:**

Take each food item and look at the nutrition information on the package. Most products will carry a label listing all the key facts about your food: the size of one serving, and the calories, fat, protein, and carbohydrates found in it, plus the levels of other key nutrients.

Work out how many grams of sugar are in the food:

- Write down how much of the food you have, and how many servings that is. It should say how many servings are in a whole box, but if it's open already, you might need to weigh what's left.

You can use the weight of a serving in grams, and divide the number of grams you have by that: I have 120 g, and one serving is 55 g, so I have 120/55 = 2.18 servings

- Then you can calculate how much sugar is in the food. Here, one serving contains 12 g sugar, so 2.18 servings would have 2.18 × 12 = 26.16 g.

Once you have a total number of grams, you could try to visualize your total as a number of 900-g (2-lb.) bags of sugar. You can divide your total by this, and imagine that many bags of sugar on the table.

**THE NUTRITION LABEL** separates out "added sugars" from naturally occurring sugars. If you want to cut down on sugar, a good way is to look for products with no added sugar; they'll taste good because naturally occurring sugars are delicious!

## WHAT IF IT DOESN'T HAVE A LABEL?

Not all products will have a nutrition label. If you're looking at fresh fruit and vegetables, or food that you've cooked at home or taken out of the package, you won't be able to just look up this information. You could use a kitchen scale to see how much you have, or look up the recipe used to cook the food, then search online for the nutrition values for equivalent products. Searching for "How much sugar is in an apple?" returns 10 g, as per the USDA (US Department of Agriculture) database.

# **LEARN ABOUT:** HOW MUCH ENERGY IS IN YOUR FOOD?

**Do you know how much energy is in different types of food? Foods high in sugars and fats will contain the most calories; healthy foods that are low in sugar and with lots of fiber will contain fewer calories.**

Can you put these food items in order, from most to least calories? Use the information on the previous pages to help you.

1. One medium apple
2. 20 g (1 oz.) plain salted potato chips
3. One T-bone steak
4. One 500-g (1-lb.) bag of chocolate candies
5. One entire 1-kg (36-oz.) chocolate cake
6. One large egg, scrambled
7. One cup of chopped celery
8. One fast food hamburger, regular fries, and regular coke
9. One clove of garlic
10. One cup of milk
11. One slice of pepperoni pizza
12. Two teaspoons of yellow mustard
13. One jelly donut
14. One medium onion
15. One glass of water
16. One plain cracker

## HOW MANY CALORIES ARE IN A WHOLE COW?

Given the information below, can you work out how many calories you'd consume if you ate a whole cow, leaving only the bones behind?

- The weight of an average cow: around 800 kg (1,764 lb.)
- The proportion of a cow's weight that's fat: 20%
- The proportion that's muscle: 55%
- The proportion that's bones: 25%
- Calories per gram in beef fat: 9
- Calories per gram in beef muscle: about 2.8

If your recommended daily consumption is 2,000 cal per day, how long could you survive eating only the cow? (Assuming you have a way to keep the meat fresh!)

**CHECK OUT** page 136 to test your cow-related knowledge!

# **LEARN ABOUT:** DANCING FOR YOUR DINNER

**How many calories would you burn with a half hour of skateboarding? How much energy would two scrambled eggs provide?**

Using the information on pages 98–99 and 104–105, match up each activity below with the food (and its calories) the activity would burn off, for an average 68-kg (150-lb.) person.

- **A.** Skateboarding for 32 minutes
- **B.** Standing still for 48 minutes
- **C.** Sleeping for 9 minutes
- **D.** Playing volleyball for a little over 4 hours
- **E.** Running up the stairs for just over 5 hours
- **F.** Laughing for 173 minutes
- **G.** Walking at a moderate pace for 61 minutes

1. 225 g (8 oz.) of salted potato chips
2. One jelly donut
3. One medium apple
4. Two scrambled eggs
5. One plain cracker
6. One teaspoon of yellow mustard
7. A whole eight-slice pepperoni pizza

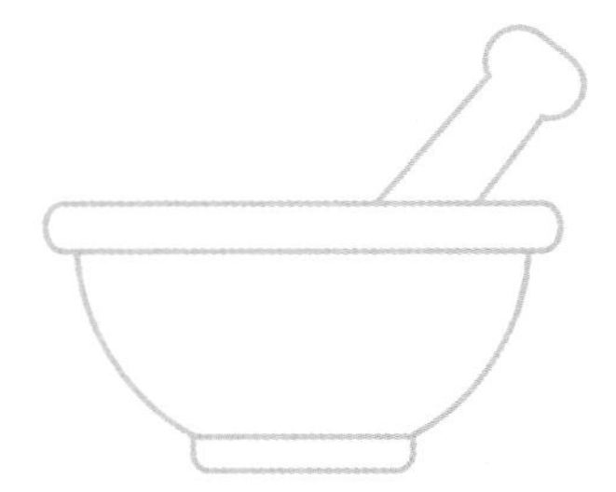

# DISCOVER: HOW DOES TASTE WORK?

**Taste is an important aspect of eating food, and it's intricately linked to our biology. The senses of taste and smell are directly connected to part of the nervous system that acts involuntarily, which is why if we smell or taste something gross, it can make us spontaneously vomit, and why tastes and smells can provoke emotional responses.**

Humans have evolved to use the taste of food as an indicator of whether or not it's good to eat. Bitter-tasting foods could potentially be poisonous, while salty or sweet-tasting foods are typically good sources of nutrients and energy.

The way food tastes depends on a variety of factors—besides the actual flavor, the texture, smell, and temperature can all affect how it tastes. This is why, if you have a cold, food tastes blander, because you can't smell it.

The tongue is a set of eight interconnected muscles used to manipulate food, assist in chewing, and push it to the back of your mouth to swallow. But it's also covered in papillae—lumps that you can see in the mirror. They give the tongue a much larger surface area than it would if it were flat, and they're all covered in taste receptors called taste buds.

On the front part of the tongue, there are 3–5 taste buds per papillus, and around 200–400 papillae. At the very back, there's a V-shaped section with larger, raised papillae, each containing thousands of taste buds. Down the sides of the tongue, toward the back, there are a series of folds, each containing several hundred taste buds. Some taste buds and taste-detecting cells are also located elsewhere in your mouth: the back of the throat, inside the nose, and even at the top of the esophagus.

## TALL TONGUE TALES

People sometimes claim that different parts of the tongue contain taste buds for detecting different flavors. This isn't true! There are different types of taste buds to detect different flavors, but they're all spread equally around the tongue, with more taste buds toward the outside edges and fewer down the middle.

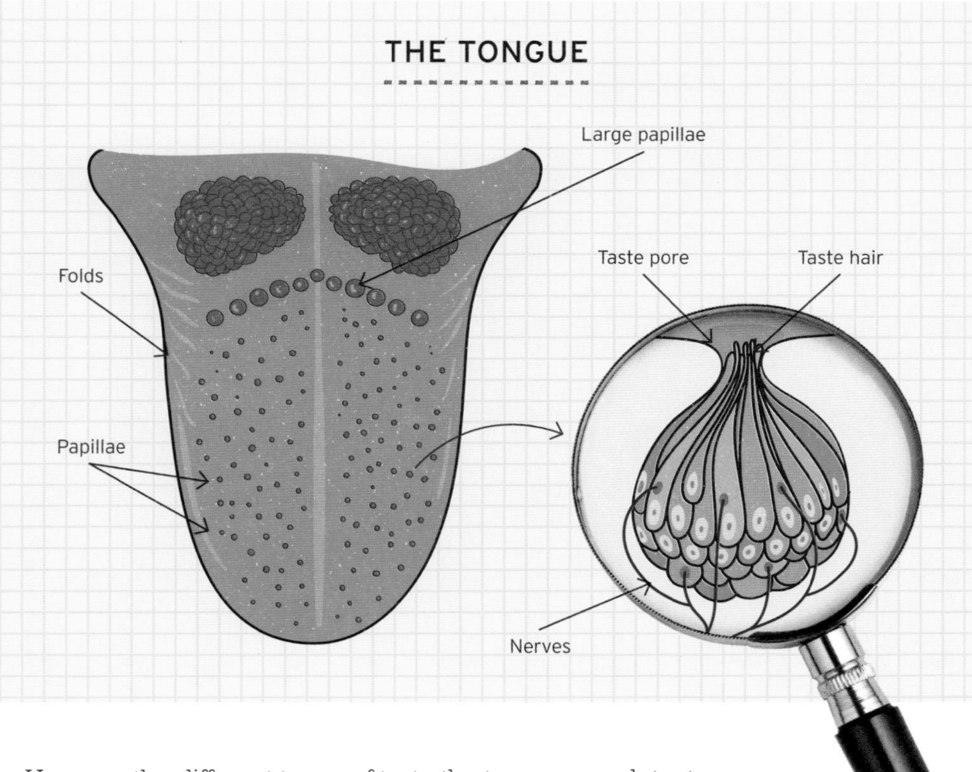


Here are the different types of taste the tongue can detect:

- **SWEET:** mainly from sugars, but also activated by some amino acids and alcohols.
- **SOUR:** activated by acidic substances such as lemon juice or vinegar.
- **SALTY:** activated by table salt (sodium chloride) or minerals like potassium or magnesium salts.
- **BITTER:** many substances cause the bitter taste buds to activate, and they're important for recognizing poisonous plants.
- **SAVORY:** sometimes also called umami, and activated by amino acids like glutamic acid or aspartic acid, which give a meaty taste.

# **LEARN ABOUT:** THE TONGUE

**It's always there, in your mouth, and it's a crucial part of eating and tasting things, but how much do you really know about your tongue? Test your knowledge below.**

## POP QUIZ: THE TONGUE

**1.** How many muscles make up the tongue?
a) 1
b) 2
c) 4
d) 8
e) 16

**2.** True or false: your tongue is unique, like a fingerprint.
a) True
b) False

**3.** How much does a human tongue weigh?
a) Around 20 g
b) Around 50 g
c) Around 70 g
d) Around 100 g
e) Around 200 g

**4.** What are the 5 tastes that you can detect?
a) Savory, sweet, sour, good, evil
b) Angry, sleepy, grumpy, happy, dopey
c) Bitter, sweet, salty, sour, spicy
d) Sour, sweet, bitter, brown, savory
e) Salty, sweet, bitter, sour, savory

**5.** Which part of the tongue can taste sweet foods?
a) The tip
b) The very back
c) Down the sides
d) In the middle
e) All of it

**6.** True or false: all your taste buds are found on the tongue.
a) True
b) False

**7.** How many taste buds are there in the average human mouth?
a) 100-200
b) 500-1000
c) 2000-8000
d) 10,000-20,000
e) 50,000+

**8.** What percentage of people can roll their tongue lengthways into a tube?
a) 0%
b) 10-15%
c) 20-35%
d) 65-80%
e) 85-100%

**9.** Which of these foods would taste sour?
a) Lime juice
b) White bread
c) Marshmallows
d) Potato chips
e) Steamed ham

**10.** True or false: dentists recommend you brush your tongue as well as your teeth.
a) True
b) False

# **DISCOVER:** EXTREME FOODS

**Modern food technology has let us create more extreme versions of some of our favorite foods. Here's a roundup of some superlative foods—things found in nature, or that have been artificially created by pushing the boundaries of food science.**

## HOTTEST CHILI

The Carolina Reaper, a red, gnarled pepper with a small pointy tail, is recognized by the *Guinness Book of World Records* as the hottest chili. Hotness of chilis is measured on the Scoville scale, which measures the concentration of capsaicin within the plant, and Reapers score over 2,000,000 Scoville Heat Units. That means a sample has to be diluted by 2,000,000 times before you can't taste the chili.

## LOWEST-CALORIE FOOD

Fruits and vegetables like celery, grapefruit, and lettuce are very low in calories because they're high in fiber and water. There's a theory that because foods like celery are high in cellulose they take more energy to digest than is gotten from eating them—that they're negative-calorie foods. This is not true, although they are much healthier than junk food!

## SMELLIEST FOOD

There are plenty of competitors for the stinkiest food, but there's one that probably deserves the crown: durian is a large fruit with a thorny rind on the outside, with flesh that's pale yellow, and tastes creamy and slightly of almonds.

But despite its delicate taste, the smell of the fruit is very different: it's described by some as being like rotten onions, turpentine, or raw sewage. It's so bad that in parts of Southeast Asia, people are banned from eating the fruit in hotels and on public transportation.

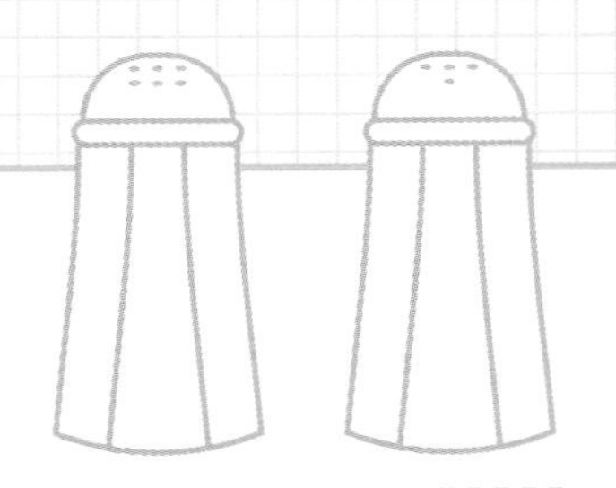

# EXPERIMENT: TASTE EXPERIMENTS

**Your sense of taste is incredibly complex, and how something tastes can be affected by its smell, texture, temperature, and even the way it looks. This experiment looks at how easily you can distinguish between different tastes when you can't see or smell the food, and when there is no difference in texture.**

You'll need a friend to pass you things when you're blindfolded!

### YOU WILL NEED:

- **Blindfold and nose clip**
- **Paper ketchup cups; forks for picking up foods**
- **Samples of foods with different flavors but similar texture:**
  - **Liquids: sugar water (sweet), salt water (salty), lemon juice (sour), tonic water (bitter)**
  - **Similar-textured foods, cut into small cubes: apple, potato, sweet potato, pear (not too ripe), onion, jicama**
  - **Strong-smelling foods, with a similar amount of each in a ketchup cup: chocolate, garlic powder, smelly cheese, powdered or fresh ginger, cinnamon crackers, garlic crackers**

**ADULT SUPERVISION REQUIRED**

Don't worry if you can't find all of these! You don't need all of them, and you can add in anything else you think will work. If you have food allergies, don't taste anything that might cause a reaction! (Check out pages 122–123 to find out more about allergies.)

### WHAT TO DO:

Design your own experiment! Make sure you only change one factor at a time, so you actually test whether you can identify the flavor without looking or smelling. Use the blindfold if you don't want to be able to see, and the nose clip if you don't want to be able to smell. Can you identify the similarly textured foods by just their taste, while wearing the nose clip and blindfold, or the strong-smelling foods just by their smell, while wearing only the blindfold and not putting them in your mouth?

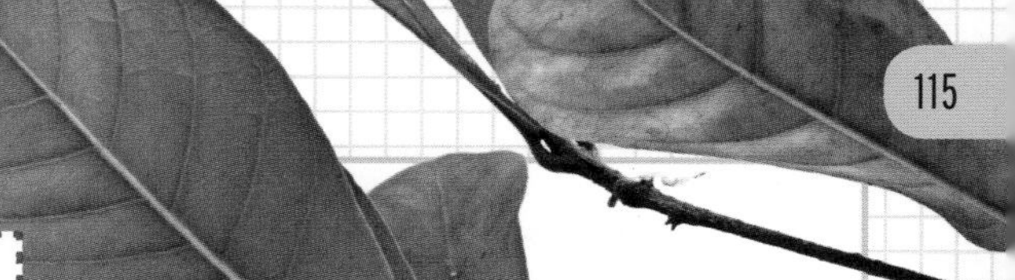

## TASTING COLORS

You could also investigate how the color of your food affects how it tastes. Try tinting the same flavor of clear soda with different colors of food dye and giving samples to someone to taste—do they think they're different flavors? Can you identify the flavor of a jelly bean with your eyes closed?

## MIRACLE BERRY TABLETS

*Synsepalum dulcificum* is a plant native to West Africa with an interesting property. It's one of a few species whose berries contain a chemical called miraculin, which affects human taste buds in a strange way. While the berries themselves don't taste sweet, the miraculin binds with the taste receptors for sour flavors, so that they register sour foods as tasting sweet.

This means if you eat something that contains miraculin, then eat a lemon, it won't taste sour as it would normally, and instead tastes sweet. The effect lasts for around half an hour before wearing off, as the miraculin is washed away by saliva. The berries have traditionally been used in cooking to sweeten the taste of foods, and Miracle Berry or Miracle Fruit tablets are sold in some countries.

# DISCOVER: HOW DO YOU SMELL?

**Your sense of smell is incredibly important and useful. It has evolved to help you survive—to detect when food is bad, and to avoid dangers such as predators or fire.**

Smell is a powerful sense, and some consider it to be more important than sight or hearing. It's so important that fully 5% of your DNA is devoted to olfaction—your ability to smell.

## HOW SMELL WORKS

Smells are made up of molecules of the substance you're smelling, which have evaporated into the air and entered your nose. This means more volatile (easily evaporated) substances smell stronger. Substances such as metals, on the other hand—whose molecules tend to stay put—don't

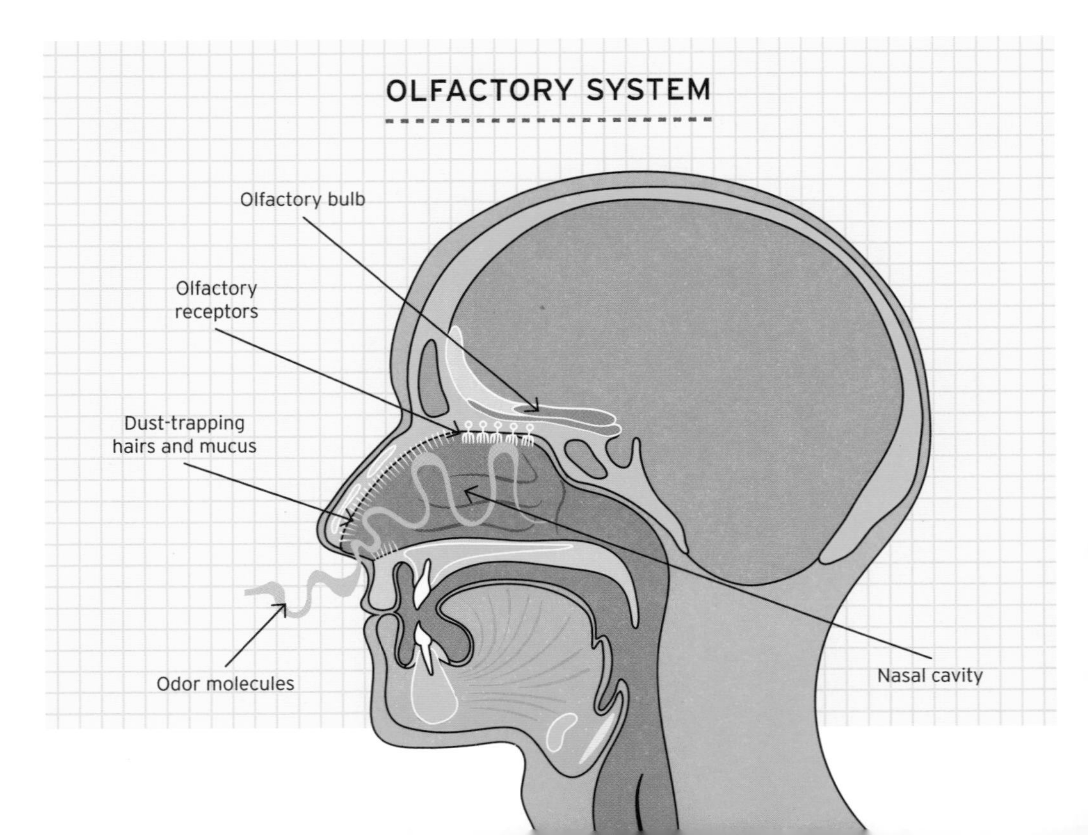

smell strongly of anything, unless you get close. Food smells better when it's hot, due to increased evaporation.

Molecules of odor pass into your nose, past hairs and mucus designed to trap dust and dirt, and hit a patch the size of a large postage stamp on the back of your nasal cavity. This area is covered in cilia—tiny hairlike projections that increase its surface area—and contains millions of olfactory receptors. Humans have around 12 million receptors and use them to distinguish between different smells.

The receptors are also covered in mucus, which traps the odor molecules so they can be analyzed. Different combinations of receptors are activated by different odors, and messages about what you're smelling are sent through nerves to the olfactory bulb, the part of the brain that interprets smells. Your nose doesn't need a lot of the substance to enter it to register a smell—you can detect a skunk with as little as 0.000000000007 g of scent—but the more of something there is, the stronger its smell will be.

## SMELL AND TASTE

The taste of food depends on a lot of things, but it's thought that 80% of the taste of our food is from its smell. The olfactory receptors in the nose will detect the smell of the food before and as you put it in your mouth, and this information is combined in the brain with taste information from your taste buds. When we're hungry, our sense of smell grows stronger.

Humans can distinguish a staggering number of different smells—at least 10,000 have been identified, but some scientists suspect it's a much higher number. Many animals, and dogs in particular, have an even better sense of smell: a dog can have as many as 200–300 million olfactory receptors. Other creatures use their sense of smell to track each other, following strong-smelling chemicals called pheromones that are released by other animals. Some species of moths can track each other over distances of up to 5 miles.

Your sense of smell can be disrupted. Simply having a cold leads to an increased amount of mucus in the nose, which can block odor molecules from getting to the receptors. Your ability to smell can be affected by many other things including nerve damage, smoking, dental problems, exposure to strong chemicals such as insecticides or solvents, radiation treatments for cancer, and nervous system disorders like Parkinson's or Alzheimer's.

# **DISCOVER:** SMELLY SCIENCE

**The molecules that evaporate from things around you and create an olfactory response in your nose come in many types, but some of the most common ones in food are esters.**

These are organic molecules, which means they're made from carbon, hydrogen, and oxygen, and they're responsible for the smells (and tastes) of many fruits and flowers.

Esters are made by combining an acid and an alcohol, and many esters are responsible for the specific smells of plants and fruit. Some common examples include:

| ALCOHOL | ACID | ESTER | SMELLS LIKE |
|---|---|---|---|
| Octanol | Acetic acid | Octyl acetate | Oranges |
| Isopentyl alcohol | Acetic acid | Isopentyl acetate | Bananas |
| Ethanol | Butyric acid | Ethyl butyrate | Pineapples |
| Ethanol | Methanoic acid | Ethyl methanoate | Raspberries |

Peaches, grapefruit, apples, plums, cherries, strawberries, raspberries, grapes, and coconut owe their smells to esters. There are also esters responsible for the smells of nuts, wood, mint, butter, oil, mushrooms, hay, glue, and spices like cinnamon and aniseed.

Once scientists worked out that the reason for smells was specific chemicals, and that they could isolate which chemicals give which fragrances, they began to figure out how to synthesize artificial esters. They created the right combinations of acids and alcohols, using concentrated sulfuric acid as a catalyst to speed up the reaction and encourage the molecules to combine.

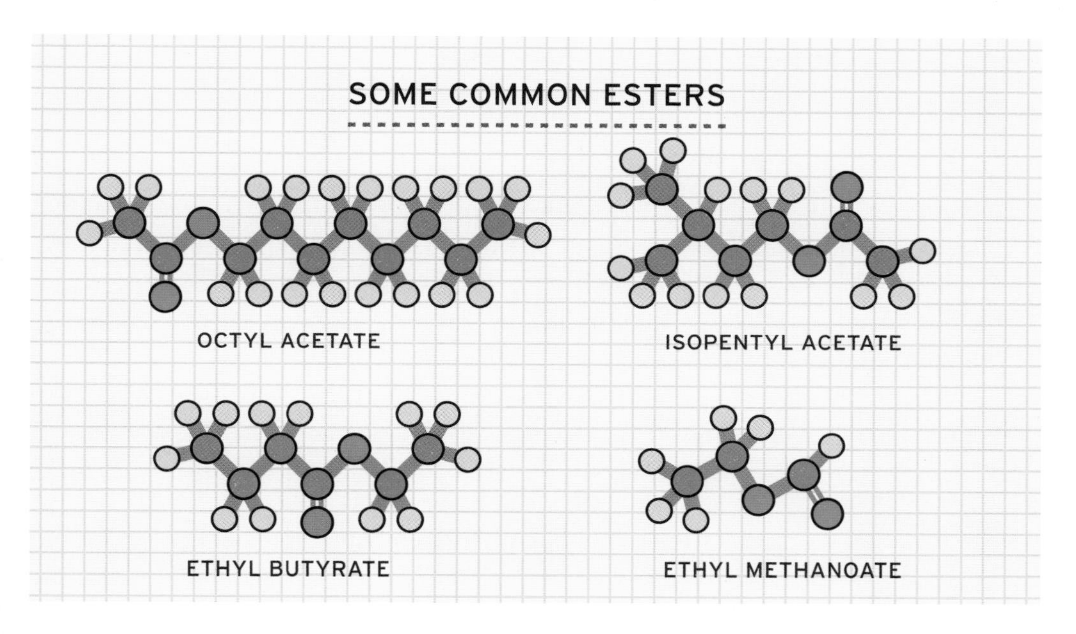


## NOT QUITE THE REAL THING

Artificial esters are widely used in synthetic flavors for candy and other foods, as well as for perfumes, air fresheners, and cosmetics. But they're sometimes not quite right—banana candy doesn't taste like real bananas, as discussed on page 11. The true experience when tasting a banana is a combination of multiple different odor molecules, along with the flavor, smell, texture, and temperature, and you can't replicate that in a candy!

Esters are produced naturally by plants to attract animals to fruit, or insects to nectar. They're also used elsewhere in nature: isopentyl acetate, which smells like bananas, is also a pheromone used by honeybees as an alarm, signaling an approaching danger.

### FUN FACT

The same type of chemical bond present in esters—the joining of an acid to an alcohol—is also present in fat molecules, whereby glycerol (an alcohol) is joined to a fatty acid.

LIME FLAVOR

BLUEBERRY FLAVOR

ORANGE FLAVOR

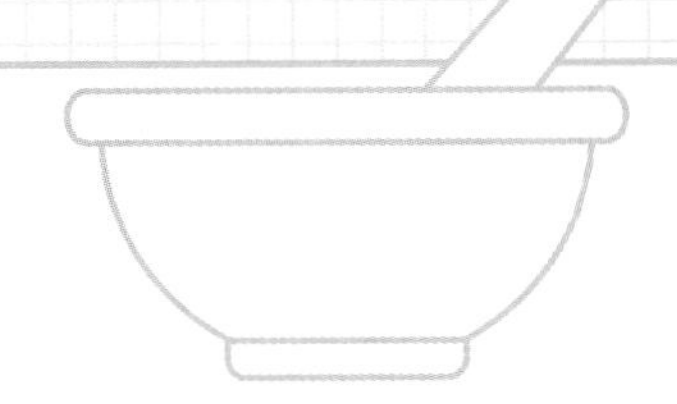

# EXPERIMENT: MAKING ORANGE PERFUME

**Among the esters that make oranges smell the way they do is octyl acetate. These esters are produced by the orange plant in order to attract animals to its fruit, but can also be used to make a perfume.**

### YOU WILL NEED:

- 1 orange
- 225 ml (7½ fl. oz.) rubbing alcohol
- Fruit zester or grater with small holes
- Bottle with a lid or a mason jar
- Colander or strainer

### WHAT TO DO:

**1.** Zest the orange. This is a cooking term, which means to grate the orange peel off the outside, but stop when you reach the white rind. (Ask an adult to help.) The peel is the source of the oils that make the orange smell, so it's the only part you need. Keep zesting until you've taken all the orange off the outside and have a white fruit.

**2.** Place the orange zest in the bottle/ jar with the alcohol, and store in a cool, dark place for 2–6 weeks. Shake the jar once or twice each day to make sure the different parts of orange zest are exposed to the alcohol, and to mix the perfume.

**3.** Pour the finished perfume through a colander or strainer to remove the orange pieces. You can put it in a perfume bottle or other lidded bottle, and dab a small amount of it on your wrists and neck, or behind your ears. The alcohol will evaporate due to your body heat, and the molecules of orange odor will be released into the air for people to smell.

## MAKE IT YOUR WAY

One way to customize your perfume is to add other strong-smelling ingredients along with the orange zest. You could add a cinnamon stick, a vanilla bean, some cloves, or a handful of cardamom pods—cut the vanilla into small pieces with a sharp knife (ask an adult to help), and crush the other spices using a mortar and pestle until they're a fine powder, and add this to your alcohol with the orange zest. Try mixing orange with other fruit zests, like lime or grapefruit. Some things will work better than others, and science is all about experimenting!

### ZESTED ORANGES

The orange peel is not only the part of the orange that contains the oils; it also forms a waterproof skin, keeping the juiciness inside. The rind is permeable to water, so if you're planning on eating or juicing the rest of the orange, you shouldn't wait too long. A zested orange will dry out as the water inside evaporates, and the juice left inside will react with oxygen in the air, so it'll taste different. If you want it to keep longer, put it in a sealed container in the refrigerator.

# **DISCOVER:** ALLERGIES

**Your immune system fights diseases, sending antibodies to attack any bacteria or viruses that have found their way into your body. But sometimes it doesn't work the way it's supposed to; your body can react with an immune response to something that isn't actually an infection.**

Food allergies are common and can cause a severe response, with symptoms including itchiness, swelling, vomiting, diarrhea, and trouble breathing.

## FOOD ALLERGIES

Different allergies are typical in different parts of the world, but some common allergies include nuts, peanuts, eggs, milk (lactose), shellfish, soy, and wheat (gluten). Specific proteins present in these foods can bind to the antibody immunoglobulin E, which mistakes the proteins for an invading disease. This causes the release of histamine, an inflammatory compound that causes more blood to flow to the affected area, which is useful if you need to bring in lots of disease-fighting white blood cells. If you're not actually sick though, it's less useful.

Depending on the type and severity of the allergy you have, the reaction can vary from feeling itchy or queasy to anaphylaxis, whose symptoms come on very rapidly. These can include a rash, vomiting, lightheadedness, and swelling of the affected part of the body. If this happens in the mouth or throat, it can cause the throat to close up, resulting in trouble breathing.

Anaphylaxis can be treated with an injection of epinephrine (adrenaline), and allergy sufferers often carry a pen-shaped epinephrine syringe, which they can use in an emergency.

## COMMON ALLERGIES

Gluten is a protein present in wheat, and sufferers of celiac disease have an allergy to it. The reaction takes place in the intestines, causing diarrhea, swelling of the abdomen, and loss of appetite. It also affects the intestine's ability to absorb nutrients, leading to undernourishment, and child sufferers might find their growth affected.

Other types of food intolerance are due to the body's inability to process certain substances. For example, lactose (the sugar in milk) is digested by an enzyme called lactase, and if your body doesn't produce enough lactase, you might need to avoid eating foods containing lactose. Lactose intolerance has symptoms including bloating, stomach pain, nausea, and diarrhea.

While an epinephrine syringe will help with immediate symptoms, for many food allergies and intolerances the only real way to avoid symptoms is to cut the foods out of your diet entirely. This can be a challenge, but it's much easier now than it has been in the past.

Awareness of food allergies has increased hugely in recent years. Many lactose-free or milk-substitute products are available, and gluten-free bread, pasta, and flour are increasingly found in stores and restaurants. The Food Allergen Labeling Consumer Protection Act requires all food sold in packaging to be labeled with eight common allergens (milk, egg, fish, shellfish, nuts, wheat, peanuts, and soybeans), if any amount of them is present.

# CHAPTER 4
# THE WORLD

DISCOVER...

LEARN...

EXPERIMENT...

# **DISCOVER:** THE ENVIRONMENTAL IMPACT OF YOUR FOOD

**Almost everything you do or use, whether traveling, eating food, or watching TV, uses energy, and therefore has an impact on the environment. It might only be a small impact, but since you're one of billions of people in the world, it adds up!**

Greenhouse gases (see the box on page 127) are produced by human activity and float up into the atmosphere of the planet, creating a layer of gas which traps the sun's heat, just like in a greenhouse. This is called the greenhouse effect, and it contributes to a process called global warming, meaning the temperature of the planet is increasing, which will result in major changes to the planet's delicate systems of weather and ecology.

While carbon dioxide has always been present in the atmosphere, human activity is increasing the levels rapidly—they're now nearly double what they were before the Industrial Revolution of the 18th–19th centuries. This could lead to irreversible changes in the climate of our planet, unless drastic action is taken.

Global warming can destroy polar bears' habitats.

## LIFESTYLE CHOICES

Since practically all our actions create some greenhouse gases, one way to slow global warming is to think about how much impact our lifestyles have, and to choose lower-impact alternatives. Decisions made on a large scale by governments and big companies will have the greatest effect overall, but there are still ways individuals can make a difference.

For example, over a quarter of the greenhouse gases produced by humans come from food and agriculture. Of that amount, almost two-thirds are from animal products, specifically from the energy and resources used in farming/processing them, and in the gases produced when livestock create methane gas using their gut.

## TWEAKING THE MENU

Thinking about what kinds of foods you're eating can have an impact too. Avocados generate about six times as much greenhouse gas as the same quantity of apples. Milk from dairy cows generates three times as much as soy milk (although not all nondairy milks are as environmentally friendly: almond milk, for instance, requires huge amounts of water to produce, which can impact the environment in other ways), and while a serving of beef or lamb generates 8–10 kg (18–22 lb.) of greenhouse gases, equivalent servings of protein from chicken, fish, pork, or cheese each produce less than a quarter as much (but still more than plant protein).

Carbon footprint calculators work out how much different activities contribute to the amount of carbon dioxide and other greenhouse gases produced. How much could be saved by switching beef for chicken, or walking instead of driving?

### GREENHOUSE GASES INCLUDE:

- carbon dioxide, which is produced when we burn fuel or waste, or cut down trees;
- methane and nitrous oxide, which are produced by agricultural activity and by the production of fossil fuels;
- fluorocarbons, which are released in smaller quantities by industrial processes but have a much stronger warming effect.

# LEARN ABOUT: FISH

**There are over 32,000 known types of fish, and many of them are eaten as food. This includes fish from the sea, freshwater fish, shellfish, and even cephalopods, such as squid and octopus. How much do you know about fish and how they can be an important part of your diet? Try the pop quiz to test your knowledge.**

## POP QUIZ: WHAT ARE FISH?

In taxonomic terms, fish are animals that live in water, breathe through gills, have limbs (fins) but no fingers or toes, and have a skull. Two of the main types of fish are cartilaginous (such as sharks and rays) and bony (including many of the common fish eaten for food, and many tropical fish kept as pets).

**1.** Which type of nutrition will eating fish contribute to your diet?
a) Protein
b) Vitamins and minerals
c) Omega-3 fatty acids
d) All of the above

**2.** Which of these statements about fish in your diet is true?
a) Since fish live in water, eating one fish is the equivalent of drinking four glasses of water
b) A balanced diet should include at least two 150-g servings of fish per week, including one of oily fish
c) Eating only fish and no other food will constitute a healthy diet
d) Fried fish is much healthier than fish that's been steamed, baked, or grilled

**3.** What name is given to someone who doesn't eat meat, but does eat fish and seafood?
a) Kangatarian
b) Pescatarian
c) Forgetful vegetarian
d) Ovovegetarian

**4.** Where does respiration occur in a fish's body?
a) In the gills
b) In the lungs
c) In the gills, heart, liver, and lungs
d) In every cell

**5.** What percentage of fish eaten in the world is farmed, instead of being caught?
a) 25%
b) 50%
c) 75%
d) 100%

**6.** According to the American Heart Association, how often should you eat a serving of fish as part of a healthy diet?
a) Once a minute
b) At least once a day
c) At least twice a week
d) At most once every two weeks

**OILY FISH** are a good source of omega-3 fatty acids, a type of fatty acid with long-chain molecules that can help to keep your heart healthy. Oily fish, which include anchovies, carp, herring, mackerel, and salmon, spend a lot more of their time swimming against a current, so their energy, stored in the form of oil, is distributed throughout their body so it can be easily burned.

# **LEARN ABOUT:** EATING FOR THE PLANET

## Which choices will help reduce your impact on the planet?

For each of the options below, choose which one you think would be a better choice for the environment. If you're not sure, think about what's involved in producing each foodstuff and how those processes might have an impact.

1. a) Meat stir-fry OR b) Tofu stir-fry
2. a) Quarter-pound burger patty (113 g; 4 oz.) OR b) 85-g (3-oz.) burger patty
3. a) Bagels from grain farmed on land which used to be rain forest OR b) Bagels from grain farmed on land which has always been field
4. a) Locally caught fish OR b) Fresh fish flown in from another country
5. a) Beef from cows fed on grain OR b) Beef from cows fed on grass
6. a) Vegetables grown with chemical fertilizers OR b) Vegetables fertilized using animal manure
7. a) Glass of soy milk OR b) Glass of almond milk
8. a) Snack that comes in a plastic wrapper OR b) Snack that doesn't need packaging (like an apple!)
9. a) Chicken kebab OR b) Lamb kebab
10. a) Plastic bottle of water OR b) Glass of filtered tap water
11. a) Ordering a large pizza and throwing half in the trash OR b) Ordering a small pizza and eating it all
12. a) Bottle of juice that's "from concentrate" OR b) Bottle of freshly squeezed juice

# **DISCOVER:** CROP ROTATION

**Growing food crops is difficult. How well the plants grow and how much food they produce can depend on many difficult-to-control factors, like the weather, insects, and weeds.**

Growing plants uses up nutrients from the soil and can cause damage to the soil structure. It can also cause a buildup of pests and disease. Crop rotation is a method that both farmers and people growing food crops in their garden can use to improve yields and the health of their plants.

The idea is to select a range of different crops to grow on different plots of land, each of which use different quantities of specific nutrients and have different insect pests and diseases. Each season, when the crops are harvested, the different types of plant—such as vegetables, grain crops, legumes, and grasses—are rotated between the different plots, and some are left unplanted or fallow. This means that if one plant uses up a lot of one particular mineral from the soil, the land has a chance to recover, and the next plant that grows there will still be able to thrive.

For example, someone growing vegetables could categorize them into brassicas, legumes, onions, potatoes, and roots, then lay out their garden so that they move the different crops around each growing season. Careful planning is needed: which plants will follow each other? Which ones need planting and harvesting at different times of year? When should fertilizers be added to replenish nutrients? Some farmers also include a crop in their rotation that animals can eat, so they have somewhere to graze livestock.

Crop rotation has been used since as long ago as 6000 BCE. Farmers in the Middle East would plant legumes and cereals in alternate years—they noticed it gave a better yield. Since then, the science behind crop rotation has become much better understood, and different systems can be designed to manage weed growth, maintain soil nutrient levels, or protect the soil structure depending on what the farmer needs.

# **DISCOVER:** EUTROPHICATION

**Using fertilizer can increase the yield of crops, and make sure plants have all the nutrients they need to grow properly. However, if not used correctly or carefully, fertilizers can have an effect on the environment and damage nearby ecosystems.**

Eutrophication is a process that's been identified as being caused by the use of fertilizers, and it has far-reaching and devastating effects.

The process starts when fertilizers added to the soil get washed by the rain into lakes and rivers, resulting in high levels of nutrients like nitrogen and phosphorous. This causes microscopic organisms in the water, like algae, to grow much more quickly—since it's fertilizer. But the algae are part of a delicate balance within a body of water; too much algae growing on the surface of a pond will create a thick green layer, blocking out light to the plants growing in the bottom of the pond.

Once these plants have died, and the algae then die and sink to the bottom of the pond as part of their natural life cycle, bacteria living in the water find themselves with a huge source of food. The dead plant matter and algae can all be digested, and the bacteria rapidly multiply in the presence of a food source. This uses up large amounts of oxygen from the water, which means fish and other larger life forms living in the water then can't breathe, so they die. This process can't be reversed, and also affects other wildlife that eat fish from the pond. It can be catastrophic.

Planes are used to spray crops with pesticides and insecticides.

While this process sometimes does occur naturally, the addition of artificial and chemical fertilizers to the soil makes it a much more serious problem. Farmers help by using less fertilizer, and techniques like crop rotation can make a difference. It's also possible to try to remove nitrogen and phosphorus from water systems when they are detected in high levels, by chemical means, or by introducing life forms like oyster reefs, which remove nitrogen naturally.

# **LEARN ABOUT:** CROP ROTATION

**Farmers and home gardeners use crop rotation to take care of the soil and make sure their plants get enough nutrients. Help design a crop rotation scheme by following the advice given below.**

## TYPES OF CROPS

**Row crops:** These include vegetables with shallow roots. There are gaps between the rows, which leave soil exposed and vulnerable to erosion. While they're profitable to grow and sell, they deplete the soil of nutrients, reduce soil quality, and damage the structure.

**Nitrogen-fixing crops:** These are legumes like clover, alfalfa, peas, and beans, which collect nitrogen from the soil and fix it in nodules as part of their root structure. This means when the plant is harvested and the roots are left in the soil, they're broken down and release nitrogen back into the soil, restoring nutrients.

**Cover crops:** These include grasses and cereals, which cover all of the soil's area with dense root systems, restoring the soil's structure and crowding out weeds. Livestock can also graze on these crops, which enriches the soil through manure.

## STRIKING A BALANCE

It's a good idea to grow a nitrogen-fixing crop in a field before you use it to grow a crop which is likely to remove nitrogen from the soil, and to use a cover crop to restore soil structure before growing something likely to damage the structure of the soil. But this must all be balanced with the need to grow profitable crops you can sell.

This crop rotation chart has already been partly filled in. The farmer wants to grow cabbages, peas, and wheat in her three fields and would like a plan for the next three years. Each field must contain a different crop each year, and the same crop shouldn't be grown in two fields at the same time. Can you complete the chart?

FIELD A
FIELD B
FIELD C
YEAR 1
PEAS
?
?
YEAR 2
?
?
WHEAT
YEAR 3
CABBAGE
?
?

# DISCOVER: BUGS FOR DINNER

**One food source that many people wouldn't consider an option is eating insects. While it might seem kind of gross to eat bugs, it's actually a pretty common practice already—nearly 2,000 species of insects are known to be edible, and around 2 billion people regularly include insects in their diet.**

Commonly eaten insects include crickets, locusts, beetles, butterflies, termites, moths, bees, wasps, ants, grasshoppers, and cicadas. Insect farms, where insects can be raised as mini livestock from larvae, already exist in many places. Products made from insects include dyes, and insects also produce silk, honey, and resin.

Insects farmed for food can be freeze-dried to be eaten whole, or powdered to make into insect flour, and the flour can be used to make other foods, such as bread, pasta, chips, burgers, and energy bars.

## WHY BUGS?

Insects are actually a really good source of protein: they contain almost as much protein as meat, but with around 60% less saturated fat. They also provide a good source of B vitamins, and minerals like iron and zinc.

Since insects are cold-blooded, they use energy very efficiently. They don't lose any energy as heat, so more of what they consume gets converted into protein. With a much smaller amount of food to feed to the animals, you can produce the same amount of edible protein; crickets are around 12 times more efficient than cows at producing protein, given the same amount of food.

It's definitely not something many people would consider; insects are traditionally associated with disease and dirt, and some people find the idea of eating insects disgusting. But insects raised in factories would be subject to strict hygiene standards and carefully monitored to make sure they were safe to eat.

Many countries, including China, Australia, India, Mexico, and South Africa, have long-established traditions of eating insects, but they don't just pick them up off the floor—it's only certain species, and they're eaten after being prepared and cooked.

## GREEN FLIES

Another factor to consider is that the world's population is growing rapidly, and the rate of food production isn't going to keep up. We can't just scale up existing food production, as it takes up huge amounts of land and resources, and it has a huge environmental impact.

So maybe insects could provide a solution to this problem. Farming insects requires significantly less land, water, and food than farming other protein sources, and the greenhouse emissions from the process are also much lower. We could even use the protein from insects as a food source for conventional livestock, freeing up land and resources for producing more environmentally friendly plant-based proteins for humans to eat.

All of this will require some careful scientific work. Insects aren't currently covered by many food laws, and the large-scale use of them as food is a new idea in many countries. Many aspects of their production and consumption haven't been tested or regulated. As the problem of where we're going to get our protein from becomes more urgent, maybe we'll find that bugs are the answer.

# **LEARN ABOUT:** COWS AND BULLS

**Cows are part of human history. They're a major food source, and have many uses. Cattle have been farmed, eaten, and traded since prehistoric times. But how much do you know about cows and bulls?**

## POP QUIZ: CATTLE

**1.** How many stomach compartments does a cow have?
a) 0
b) 1
c) 4
d) 7

**2.** Which of these is not a term used for a part of the cow used for meat?
a) Brisket
b) Chuck
c) Rib
d) Frank

**3.** How much of the meat consumed in the world comes from cattle?
a) 10%
b) 25%
c) 50%
d) 80%

**4.** Which of these methods are used to tenderize beef, to make it softer?
a) Leave it in the refrigerator
b) Hit it with a hammer
c) Add some pineapple
d) All of the above

**5.** Cows' eyes are on the sides of their head. How wide is their field of vision?
a) 180°
b) 240°
c) 330°
d) 360°

**6.** Which of these is not something that bone char, made from cattle bones, has been used for?
a) Eating as food
b) Protecting satellites in space from the sun's heat
c) Refining sugar, to remove impurities
d) Making paint, printing ink, and calligraphy ink

**7.** Which of the following contain products from a cow?
a) Baseball gloves
b) Gummy bears
c) Paintbrushes
d) All of the above

**8.** How big was the heaviest bull ever recorded?
a) About 600 kg (1,300 lb.) –the weight of around 260 house bricks
b) About 1,000 kg (1.1 tons) –the weight of a fully grown buffalo
c) About 1,700 kg (1.9 tons) –the weight of a large car
d) About 2,700 kg (3 tons) –the weight of a blue whale's tongue

**9.** Which of the following is not something cattle have been observed to do?
a) Tell the difference between different humans
b) Recognize their mother's voice
c) Bite grass off the ground
d) Remember the location of food sources for several hours

**10.** Which of the following parts of a cow are not eaten as food?
a) Liver and kidneys
b) Tongue
c) Udders
d) Brain
e) None of the above

# **LEARN ABOUT:** INSECTS AS FOOD AROUND THE WORLD

**Plenty of countries consume insects as food, as part of their traditional national cuisine. Can you make a guess and match each insect to a country where it's commonly eaten?**

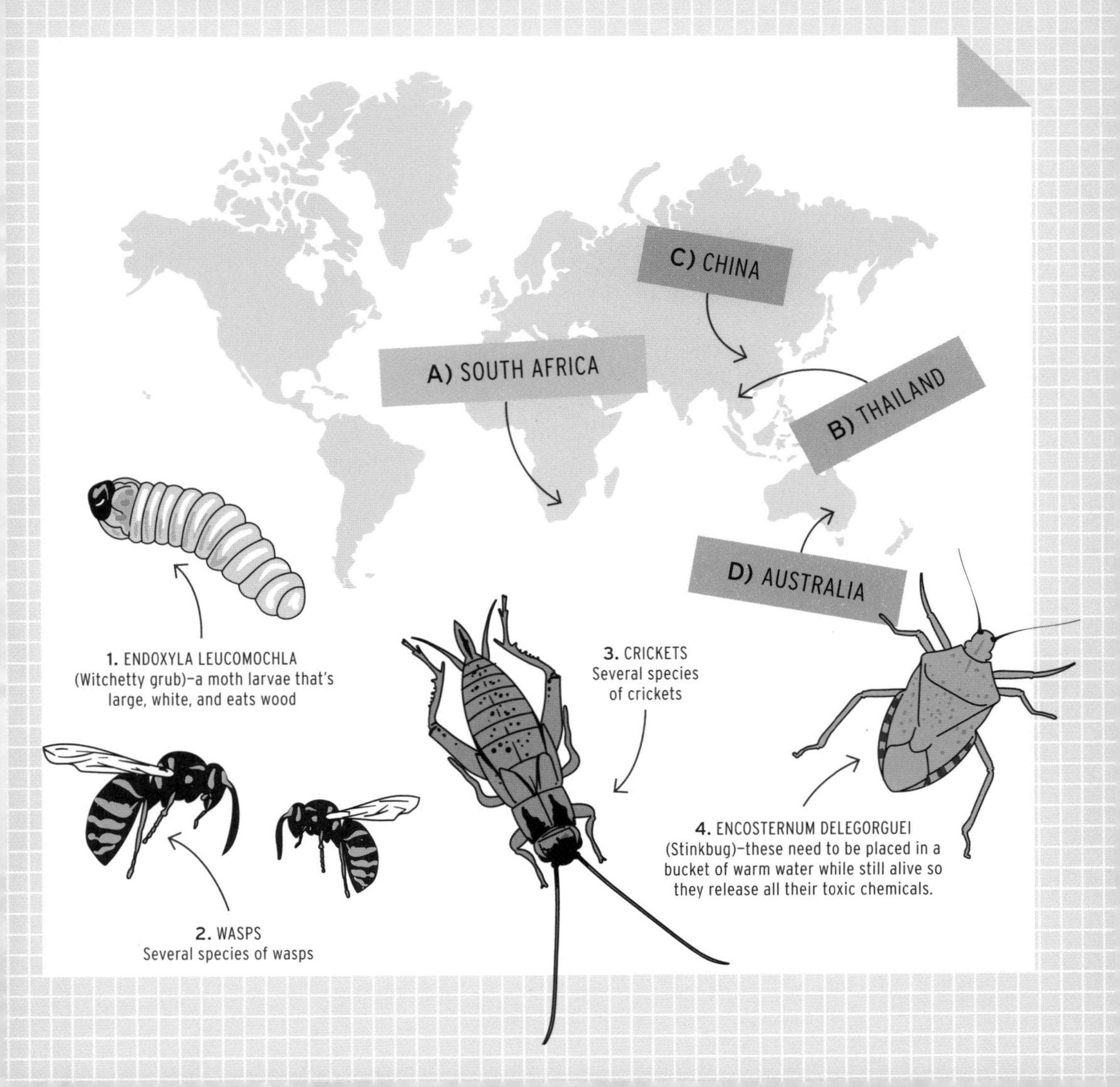

# **EXPERIMENT:** DISSECT A CHICKEN LEG

**You may have encountered a raw chicken leg before—maybe when someone was about to toss it on a grill! This experiment looks at all the parts of the chicken leg carefully, to see what it's made of and how it all connects together.**

Even though you're not a chicken, you do share around 60% of its DNA, and many of its structures and tissues are also found in your own body.

### YOU WILL NEED:

- **1 whole chicken leg, including thigh, with skin**
- **Sharp knife (ask an adult to help)**
- **Cutting board**

ADULT SUPERVISION REQUIRED

### WHAT TO DO:

**1.** Raw chicken can harbor dangerous bacteria and cause food poisoning. Before and after you handle raw meat, wash your hands thoroughly with soap and hot water.

**2.** Place the chicken on the board and examine its skin.

- Can you see where feathers have been plucked out (or are there still some attached)?
- Rub the skin. Does it feel like human skin? What if you close your eyes?

**3.** Pick up the leg and bend the knee joint carefully back and forth. It should bend just like your own knee or elbow—in one direction only, and through a limited angle.

**4.** Peel back the skin of the chicken leg and look underneath.

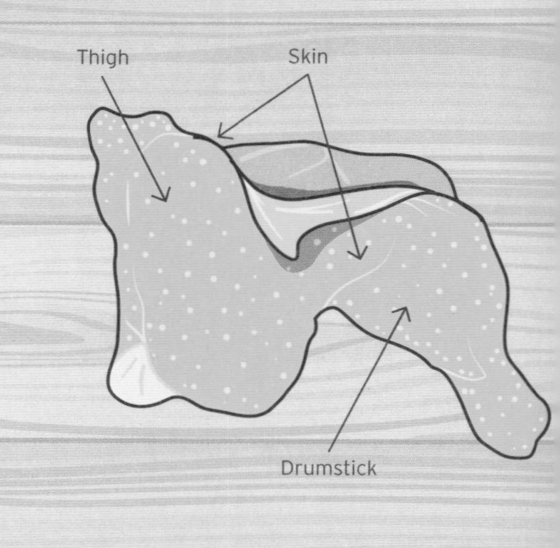

• The pink part is muscle tissue. You may also see some blood vessels running through the muscle. Muscle tissue is made from fibers of protein, which slide past each other when the muscle contracts.
• There might also be opaque white parts on the leg. These are fat deposits, used to store energy and help keep the chicken warm.

**5.** Peel the skin all the way back and take it off. You may need to use the tip of the knife to cut it away. (Ask an adult for help.)

**6.** At the ends of the muscles, you should be able to find tendons. These join the muscle to the bone.

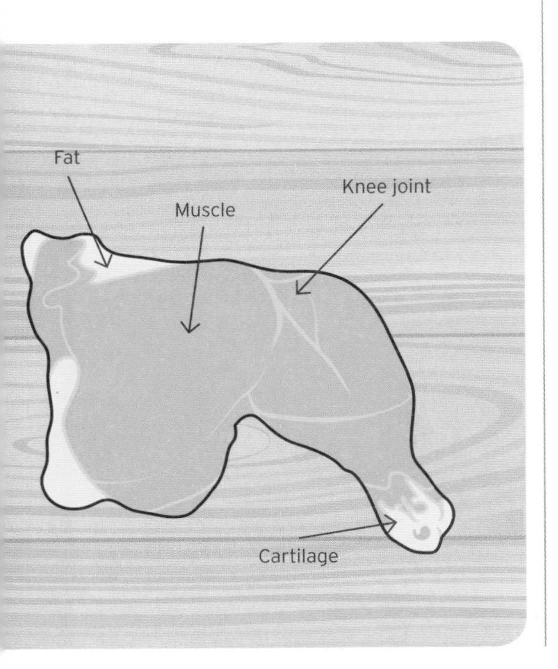


• Tendons are made from a strong fibrous protein called collagen, and they are inelastic and tough.

**7.** Where the two bones join, you may be able to find a ligament. This is much like a tendon, but it connects bones to other bones and holds the joint together.
• The ligaments in your joints stretch a little, but they do so slowly, which is why people should warm up before exercise. Making sudden, quick movements can damage the ligaments if they haven't been gently stretched first.

**8.** With an adult's help, use the knife to remove the muscle and fat tissue from the bone.

**9.** Cut through the middle of the knee joint between the two bones.
• You should be able to see cartilage on the ends of the bones—a waxy translucent substance, which protects the bones when they move over each other.
• You might see a dark-red substance inside the bones themselves. This is bone marrow; it's where new blood cells are produced.

Don't forget to wash your hands, the knife, and the cutting board thoroughly after you've finished!

# DISCOVER: GENETICS AND SELECTIVE BREEDING

**Many plants and animals reproduce by combining material from two parents. The DNA inside cells determines characteristics, and while 99.9% of human DNA is identical, the tiny amount of variation means people show variation in height, eye color, etc.**

## A DNA LOTTERY

When two organisms produce offspring together, their DNA combines to create the DNA of the child—which can take its characteristics from one of the parents, or a mixture of both. Because of the way DNA works, it's hard to predict the result, and sometimes traits occur that aren't present in either parent. For example, it's possible for two brown-eyed parents to have a blue-eyed baby.

Generally, having parents with given characteristics increases the likelihood of inheriting them. For example, if two tall people have a child, it is more likely that their child will also be taller than average.

Obviously, not many people pick their partner based on whether they want a tall child, but if you're a farmer, you can make use of this to create plants and animals closer to your ideal—trees with bigger fruit or animals that produce more milk.

## SELECTIVE BREEDING

Selective breeding is the process of choosing plants and animals to create offspring–either by manually pollinating the flowers of a plant with the right pollen, or by artificially inseminating livestock.

Find the best examples of the properties you want, breed them together, and then choose the best ones from their offspring to breed again. While most of the animals will be average, by selectively only breeding the ones that have more desirable properties, you can change the way an animal or plant looks, or produces food.

This process has been used since early prehistoric times and has created most of the food plants and livestock animals we use for food today.

Think about modern-day domestic dogs. They don't exist in the wild; the closest wild relative for many dogs is the wolf. Over centuries of selective

## MAIZE THREE WAYS

The modern plant we know as maize (corn) has as its ancestor a plant called teosinte, which originated in Mesoamerica (modern-day Central America) and was selectively grown to become more like the corn we eat today.

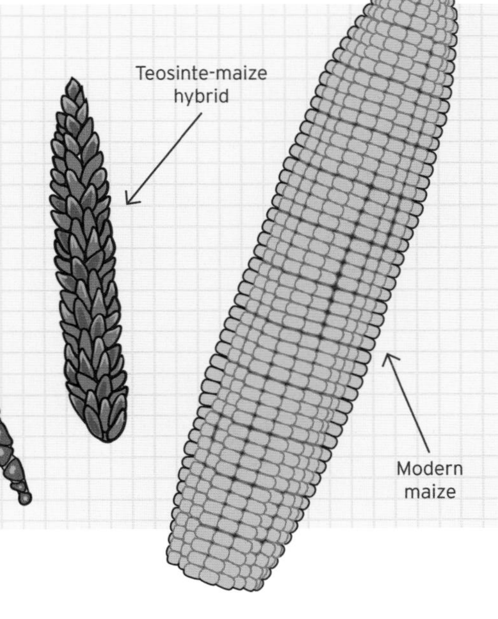


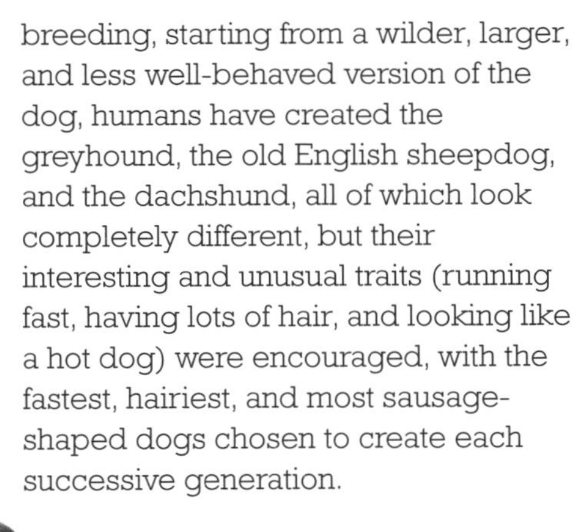

breeding, starting from a wilder, larger, and less well-behaved version of the dog, humans have created the greyhound, the old English sheepdog, and the dachshund, all of which look completely different, but their interesting and unusual traits (running fast, having lots of hair, and looking like a hot dog) were encouraged, with the fastest, hairiest, and most sausage-shaped dogs chosen to create each successive generation.

# LEARN ABOUT: NAME THAT HYBRID

**Selective breeding is used to make plants that grow bigger fruit, and animals that produce more meat, but people also breed animals to create hybrids. A species is a group of animals which can produce fertile offspring when they breed with each other, but hybrids aren't always able to reproduce themselves.**

Guess which of these animal hybrids come from which pair of species.

| | |
|---|---|
| **A)** ZEBRA + DONKEY | **1.** Dark gray cetacean (aquatic mammal) with 66 teeth |
| **B)** LION + TIGER | **2.** Small-bodied ursine with white fur and brown patches, with partly hairy soles |
| **C)** BOTTLENOSE DOLPHIN + FALSE KILLER WHALE | **3.** Largest known feline; enjoys swimming and is sociable |
| **D)** GRIZZLY BEAR + POLAR BEAR | **4.** Ungulate mammal with no humps that's large, strong, and produces soft wool |
| **E)** DOMESTIC COW + AMERICAN BISON | **5.** Small four-legged equine mammal that's patient, hardy, and intelligent |
| **F)** DONKEY + HORSE | **6.** Small equine mammal with brown torso and black-and-white striped legs |
| **G)** CAMEL + LLAMA | **7.** Bovine mammal with good cold resistance and lean, tasty meat |

Different animal pairings can result in different hybrids depending on the gender of the parent species—for example, a hinny is the offspring of a male horse and a female donkey. While most hybrids have characteristics which combine their parents' traits—for example, hybrids with one parent a zebra, which are generally termed zebroids, mostly have some black-and-white stripes somewhere on their body—sometimes hybrids can display traits which mean they're an improvement over their parents, and this is particularly useful in hybrid plants.

ZEBROID

## SOME PLANT HYBRIDS

### Olympia spinach

Olympia is a spinach hybrid which grows thicker, darker leaves and does well in high temperatures. It also has a great flavor and is resistant to disease.

### Meyer lemons

A cross between a citron (similar to a lemon) and a mandarin/pomelo hybrid (similar to an orange), Meyer lemons are sweeter and less acidic than normal lemons, and rounder with a slightly orange color.

### Hybrid maize

The process of selectively breeding modern maize plants from teosinte included creating hybrids—breeding together different species of corn—and the resulting plants had bigger, sweeter kernels.

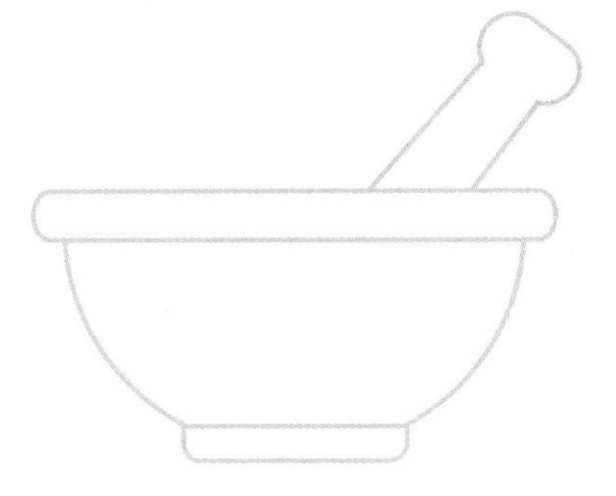

# **DISCOVER:** GENETICALLY MODIFIED FOOD

**Most food products come from genetically manipulated species. The description "genetically engineered" is given to plants and animals whose DNA has been altered by scientists in a lab, and this has produced many exciting new food crops and animals.**

## GENETIC ENGINEERING

While humans have been modifying the DNA of plants and animals for thousands of years through selective breeding, it's only in the last century that we've known the science behind it and understood what's actually happening inside the cells. The structure of DNA was discovered in 1953, and for decades scientists have worked to understand the code behind it—how the sequence of nucleotides (remember the candy DNA from pages 92–93) determines the way an organism will look and behave.

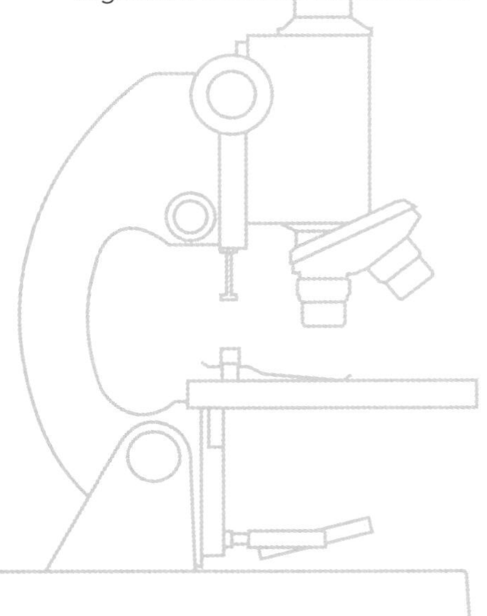

It's possible through manipulation of DNA to create new plants and animals, joining together sections of DNA from different species and placing the result in an egg/seed to be grown into a full organism. Such creatures are called genetically engineered or genetically modified (GM), and it's a technology with huge potential—creating disease-resistant plants which grow more quickly, reducing the need for pesticides and fertilizers.

## EXAMPLES OF GM PRODUCTS

Most GM crops are designed to resist diseases and pests. Papayas, plums, potatoes, zucchini, corn, and soybeans have been genetically modified to resist specific diseases and insect species and are grown all over the world—and have sometimes saved whole industries from collapse by preventing viral outbreaks.

Golden rice has been engineered to contain beta-carotene, a yellow pigment that makes the rice a golden color and which is a great source of vitamin A. It was designed to be grown in countries where diets don't

contain enough vitamin A—causing blindness and death in thousands annually—but where people consume rice as a staple crop.

Even when the goal of genetic modification isn't to save lives, it can still be useful. Arctic apples are genetically engineered to turn brown less quickly, by producing less polyphenol oxidase.

Some uses of GM are even more exciting. Animals can be genetically modified to produce medicinal proteins in their milk—GM goats have been used to produce an anticoagulant called antithrombin, for treating blood diseases.

## IS GM SCARY?

Some people find GM frightening, and scientists have to be careful to check and test GM products so that they meet strict regulations. Historically, introducing new species into an environment—for example, by trade boats—has caused problems when new, stronger species interact with existing ones. GM crops are kept isolated and grown in closed farms until they're fully tested, and they're often bred to be infertile, so they can't reproduce naturally.

It's possible that changing an organism's DNA might have different effects than those intended—there are still many aspects of DNA that we don't understand—but work on GM crops is done carefully and under strict rules, with plenty of testing and safeguards in place. The possibilities of a world with plenty of food for everyone, with less pollution from pesticides, are a good motivation to experiment!

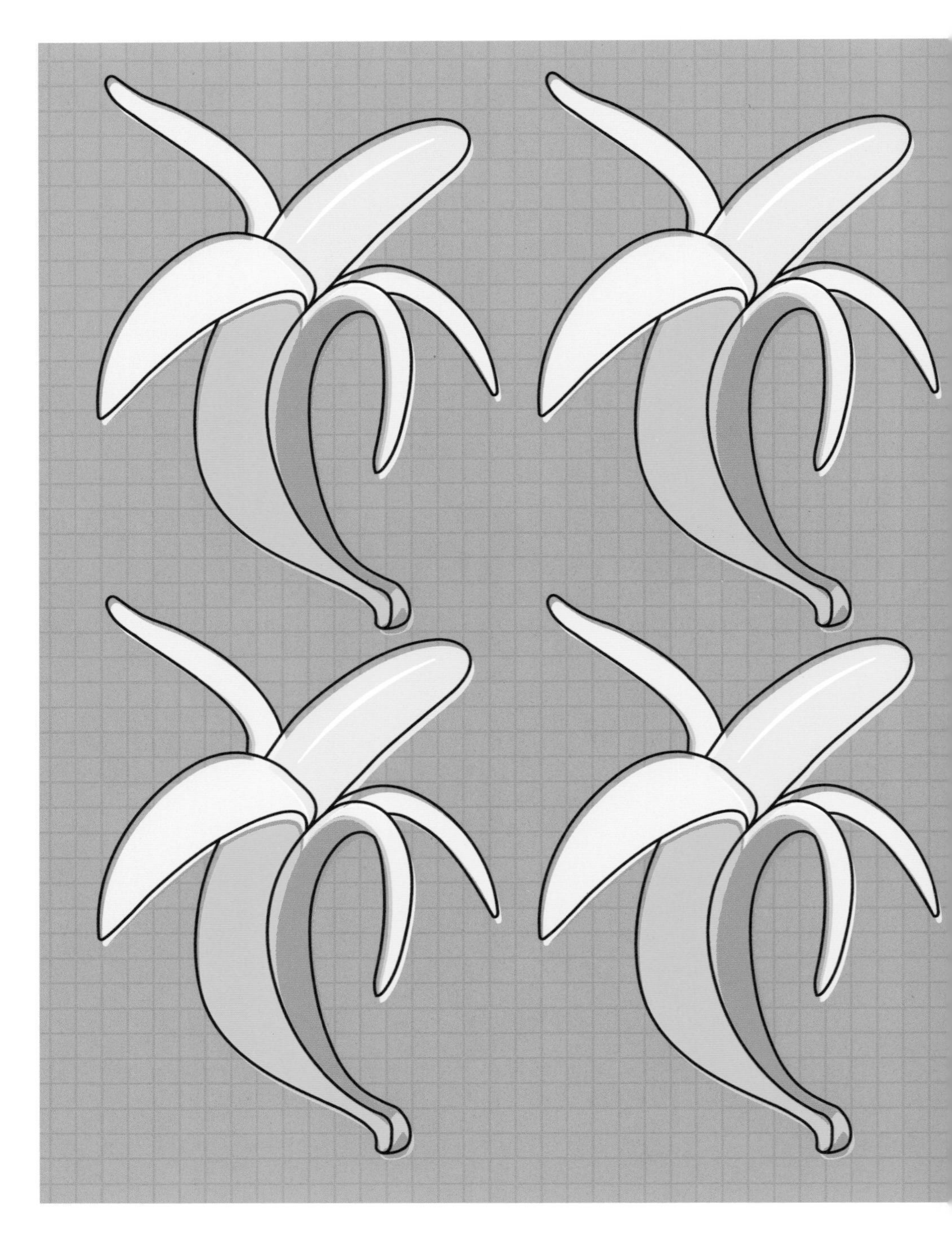

# THE ANSWERS

# ANSWERS

## CHAPTER 1 PLANTS

### P. 14 BANANAS

1. **a)** It will float. Bananas are waterproof, so it wouldn't dissolve, but they're less dense than water.

2. **b)** Smaller. We've selectively grown them to be bigger, as bigger bananas mean more tasty fruit!

3. **d)** All of these metals are present, but the one with the highest concentration is potassium. A small banana (weighing 100 g/ 3.5 oz.) contains around 360 mg (0.01 oz.) of potassium, around 27 mg of magnesium, and 0.3 mg of iron.

4. **c)** Banana leaves.

5. **a)** Hand. The bananas within a hand are referred to as fingers.

6. **b)** 50%. Humans share 50% of their DNA with a banana—and 90% with a cat!

7. **f)** All of the above.
**a)** Rubbing the inside of a banana peel on your skin has been claimed to relieve itching from insect bites and poison ivy, and to treat skin conditions including eczema.

**b)** Rubbing a banana peel on your teeth is claimed by some to whiten teeth; however, there's no evidence for this, and the American Dental Association doesn't recommend rubbing fruit peel on your teeth—fruit is often acidic, which can damage tooth enamel.

**c)** The sweetness of banana peels means they'll attract butterflies, if left in the garden—but they'll also attract flies and wasps!

**d)** Rubbing the inside of a banana peel on leather shoes or bags, then buffing them clean, can make them shinier. It also apparently works on plant leaves and silverware. It's cheaper than buying polish ... and you get to eat a banana!

**e)** It's claimed that taping a piece of banana peel over a splinter in your skin and leaving it there for a while will make it easier to remove the splinter—enzymes present in the banana can soften the surrounding tissue.

8. **b)** River carp.

9. **a)** 8. Patrick Bertoletti from Chicago, Illinois, set the record in 2012. That same day, Patrick also set the record for most cloves of garlic eaten in one minute (36) and most mini pickles eaten in one minute (16).

10. **b)** Blueberry.

### P. 15 PARTS OF A CELL

**Animal cell:** mitochondria, cell membrane, nucleus, cytoplasm, ribosomes
**Plant cell:** mitochondria, chloroplasts, cell membrane, nucleus, cell wall, cytoplasm, ribosomes, vacuole
**Yeast cell:** mitochondria, cell membrane, nucleus, cell wall, cytoplasm, ribosomes
**Bacteria cell:** cell membrane, cell wall, plasmids, cytoplasm, ribosomes

1. **Roasted ham:** animal cells
2. **Dressed salad:** plant cells, yeast cells (vinegar)
3. **Cheeseburger with lettuce:** animal cells (burger patty), yeast cells (bread), plant cells
4. **Strawberry yogurt:** bacteria cells, plant cells
5. **Chicken wings:** animal cells

### P. 23 HOW MANY IS THAT?

150,000,000,000 cells ÷ 50,000 per bag = 3,000,000 (three million) bags of rice.

If you placed all these bags of rice on the floor, they'd cover the area of more than five and a half full-size football fields!

### P. 23 SURPRISE CELLULOSE!

DO contain cellulose:

- A pencil—wood is 40-50% cellulose.
- Lettuce leaves, an apple, a stick of celery, a bell pepper, a daffodil stem—made mostly of plant cells, with cellulose in cell walls.
- $100 bill—paper, and the cotton/linen fibers used to make banknotes, contain cellulose.
- Candy wrappers—often made from cellophane.
- This book—paper is made from wood and contains cellulose.
- Milkshake—cellulose is used as a thickener.
- A pair of jeans, a cotton T-shirt—cotton is 90% cellulose, and denim is made from cotton.

DO NOT contain cellulose:

- A ham—animal cells don't have cell walls.
- Keys, a nickel, a knife and fork, a saucepan—made of metal.
- Rocks—made from minerals, not cells.
- Dinner plate—ceramic is made from minerals, too.

### P. 46 PHOTOSYNTHESIS: TRUE OR FALSE

**1.** This is false. Overall, plants take in carbon dioxide and use it to grow: the substances that make up plant cells contain a lot of carbon, and large trees can grow by up to 100 kg (220 lb.) per year, of which 38 kg (84 lb.) would be carbon. They do produce carbon dioxide by respiration, but it's not as much as they take in during photosynthesis.

**2.** This is true. Having a large number of plants on Earth's surface is good for the planet; it counterbalances the effect of humans and animals breathing oxygen in and carbon dioxide out. Deforestation (cutting down large areas of forest and plant life) and large-scale animal farming (rearing huge numbers of cows and sheep) mean this balance is harder to maintain, and if we want to look after the planet, we need to take care of our plants.

**3.** The answer to this is: we don't know! Some plants have been found in laboratory tests to remove organic substances from the air, and the microorganisms in the soil plants are potted in also play a part in cleaning the air. However, these findings haven't been successfully replicated in a real-world environment, and it's not been proven whether plants actually remove toxins from the air. But they do produce oxygen, and they look nice!

**4.** This is true on average, but the rate of photosynthesis depends on a lot of factors. Plants give off different amounts of oxygen at different phases in their growing cycle, and they take in different amounts of carbon dioxide, depending on the temperature and level of carbon dioxide in the air around them.

**5.** It's true! The plant detects that winter—a time of less sunlight—is on its way, and stops producing chlorophyll. Then the leaves dry out and fall off, so the tree can save energy.

**6.** This is false. Photosynthesis mostly occurs during the day, when the sun is shining, so plants produce less oxygen, and take in less carbon dioxide, at night.

**7.** It's true... and not true! *Elysia chlorotica* is a species of sea slug found on the East Coast of the US and Canada which eats and partially digests algae, and uses its chlorophyll to produce food from sunlight, as part of a symbiotic relationship.

**8.** This is true. The pores are called stomata—from a Greek word meaning "mouth"—and they open and close to control the amount of carbon dioxide coming in, and water evaporating out of the plant. This is useful when the weather is dry and the plant needs to conserve its water.

**9.** This is false. The chlorophyll must still be present, or else the plant would die. It's masked by stronger red and purple pigments present in the leaves, called anthocyanins. These protect the tree from strong sunlight.

**10.** This is true, and the carbon that was being stored in the trees is released as carbon dioxide when they're burned. What's more, those trees are no longer using up carbon dioxide in photosynthesis, and this can add huge amounts of $CO_2$ into the atmosphere.

### P. 47 POTATOES

**1. a)** A tuber is a part of the plant's stem that's swollen to store food and nutrients. While it is underground, it isn't technically part of the root. Potatoes are stem tubers, while sweet potatoes are a slightly different type of plant called a root tuber, whose tubers are made from swollen sections of root.

**2. d)** Sweet potatoes are from the family Convolvulaceae. In Spanish, the words for "potato" and "sweet potato"—"patata" and "batata"—are very similar, which might be why they're both called "potato" in English, even though they're different plants.

**3. b)** "Tuber" means lump, bump, or swelling.

**4. c)** 79% of potato flesh is water.

**5. b)** In ideal conditions, an average harvest is around 1-2 kg (3-5 lb.) of potatoes, which is about 5-15 potatoes per plant. This will vary, based on the variety of potatoes grown, and how well they are taken care of.

**6. c)** 4.98 kg (10 lb., 14 oz.), grown by Peter Glazebook, UK. Peter has previously also held records for the world's longest carrot, heaviest parsnip, and longest beet.

**7. c)** September. They're planted around April, and take around 15-20 weeks to grow to maturity.

**8. d)** There are around 4,000 different known varieties of potatoes, although only a small number of them are commercially available.

**9. a)** South America. Many varieties of potatoes still grow in the Andes—the mountain range that runs through Peru and Bolivia, where the first potato farmers lived. Potatoes are now grown in countries all over the world.

**10. b)** A potato on its own has about 77 calories, but the average candy bar has more like 500 calories—so you'd need about 6.5 potatoes to match one candy bar.

## CHAPTER 2 FOOD

### P. 68 BREAD AND YEAST

**1. d)** All of the above. Sugar is most commonly used for baked breads and pastries that use yeast, but boiled potatoes have historically been used as a sugar source in some breads.

**2. c)** *Saccharomyces cerevisiae* means "sugar-eating fungus."

**3. a)** Ancient Egypt. Hieroglyphics suggest that ancient Egyptians were using yeast to produce alcoholic drinks and make bread rise over 5,000 years ago.

**4. c)** Tortillas. Donuts are made with yeast, and king cakes and cinnamon buns use yeast to make them rise. Tortillas are unleavened bread, which doesn't rise.

**5. a)** Fermentation.

**6. d)** All of the above. While yeast cells feed on sugar, if the concentration of sugar around them is too high, they will lose water due to osmosis (see page 57). Salt in high concentrations is also bad for yeast for similar reasons, and adding a small amount of salt to a dough recipe can be a way to regulate yeast growth for a more even result. Alcohol is toxic to yeast in high concentrations.

**7. b)** Baking soda bread. Soda bread uses sodium bicarbonate to generate the carbon dioxide to rise. Some non-alcoholic drinks, including root beer, use yeast to produce carbon dioxide to make them fizzy, but fermentation is stopped before alcohol is produced. Plants in aquariums can be supplied with carbon dioxide from yeast, and some probiotic supplements contain yeast, and have been shown to reduce diarrhea symptoms.

**8. b)** Soak it in warm water.

**9. a)** It's often used in vegan cooking to replace the taste of cheese. It's naturally low in salt, fat, and sugar, and can be a good source of vitamins.

**10. d)** A fungus.

## P. 69 BACTERIA AND MOLD

**1.** Things which make bacteria grow more quickly: a warm environment; low salt levels; sugars, starches, and proteins present (for food!); moist surroundings.

Things which inhibit bacterial growth: dry surroundings; inside a hot oven; in a freezer; high salt levels; no sugars, starches, or proteins present.

**2.** Single-cell organism
**Bacteria:** Yes **Mold:** No

Forms long filaments
**Bacteria:** No **Mold:** Yes

Reproduces by dividing
**Bacteria:** Yes **Mold:** No

Reproduces by releasing spores
**Bacteria:** No **Mold:** Yes

Considered a microbe
**Bacteria:** Yes **Mold:** Yes

Comes in many different shapes and forms
**Bacteria:** Yes **Mold:** Yes

Causes food to spoil
**Bacteria:** Yes **Mold:** Yes

Typically wears a hat
**Bacteria:** No **Mold:** No

Can be used in making cheese
**Bacteria:** Yes **Mold:** Yes

Too small to be visible unless allowed to grow
**Bacteria:** Yes **Mold:** Yes

Cannot grow when the temperature is too low
**Bacteria:** Yes **Mold:** Yes

# CHAPTER 3 YOU

## P. 87 PARTS OF THE DIGESTIVE SYSTEM

1. Mouth
2. Liver
3. Gallbladder
4. Cecum
5. Appendix
6. Salivary glands
7. Esophagus
8. Stomach
9. Pancreas
10. Large intestine
11. Small intestine
12. Rectum
13. Anus

## P. 102 PICKING A GOOD DIET

**Diet A:** 1 No, 2 Yes, 3 Yes, 4 Yes, 5 Yes, 6 No, 7 Yes, 8 No (3,775 cal!)

**Diet B:** 1 No, 2 Yes, 3 Yes, 4 Yes, 5 No, 6 No, 7 Yes, 8 No (3,598 cal)

**Diet C:** 1 Yes, 2 No, 3 Yes, 4 Yes, 5 No, 6 Yes, 7 Yes, 8 No (way too low–1,535 cal)

**Diet D:** 1 No, 2 No, 3 Yes, 4 Yes, 5 No, 6 No, 7 No, 8 Yes (2,135 cal)

## P. 108 HOW MUCH ENERGY IS IN YOUR FOOD?

In order from most to fewest calories:

- One entire 1-kg (36-oz.) chocolate cake = 3,840 cal
- One 500-g (1-lb) bag of chocolate candies = 1,425 cal
- One fast food hamburger, regular fries, and regular coke = 1,090 cal
- One T-bone steak = 580 cal
- One slice of pepperoni pizza = 298 cal
- One jelly donut = 289 cal
- 20 g (1 oz.) salted potato chips = 155 cal
- One cup of milk = 149 cal
- One large egg, scrambled = 102 cal
- One medium apple = 72 cal
- One plain cracker = 59 cal
- One medium onion = 44 cal
- One cup of chopped celery = 18 cal
- Two teaspoons of yellow mustard = 6 cal
- One clove of garlic = 5 cal
- One glass of water = 0 cal

### P. 108 HOW MANY CALORIES ARE IN A WHOLE COW?

If an 800-kg (1,764-lb.) cow is 20% fat and 55% muscle, that means it contains around 160 kg (353 lb.) of fat and 440 kg (970 lb.) of muscle. Since 1 kg = 1,000 g, that means the total number of calories will be (160,000 x 9) + (440,000 x 2.8) = 2,672,000 cal. This is enough to survive for 1,336 days, which is more than three and a half years! You might get sick of the taste of beef before then, though.

### P. 109 DANCING FOR YOUR DINNER

A 5, B 3, C 6, D 1, E 7, F 2, G 4

### P. 112 THE TONGUE

1. **d)** 8 muscles make up the tongue, and four of them are the only muscles in the body not directly attached to your bones!

2. **a)** True.

3. **c)** 70 g—measured from the epiglottis (at the top of the esophagus) to the tip, the average tongue is 9 cm (3.5 in.) long.

4. **e)** Salty, sweet, bitter, sour, savory.

5. **e)** All of it—there are sweet-sensing taste buds on all parts of the tongue.

6. **b)** Especially in young children, taste buds and taste-sensing cells are found all over the mouth.

7. **c)** 2000-8000.

8. **d)** 65-80%. It was thought to be a genetic trait, but there's evidence to suggest it can be learned.

9. **a)** Lime juice.

10. **a)** True. Bacteria collecting on the tongue can cause bad breath, and brushing your tongue, along with brushing and flossing your teeth, forms part of a good hygiene routine.

## CHAPTER 4 THE WORLD

### P. 128 WHAT ARE FISH?

1. **d)** All of the above.

2. **b)** A diet needs to be balanced. Frying food adds oil or fat, which makes it less healthy.

3. **b)** Pescatarian. An ovovegetarian follows a vegetarian diet that includes eggs but not dairy, and a kangatarian will eat only kangaroo meat in addition to a vegetarian diet.

4. **d)** Respiration occurs in all living cells, to produce energy.

5. **b)** 50%. Farmed fish are raised in tanks or enclosures, and commonly farmed species include salmon, tuna, cod, trout, and halibut.

6. **c)** It contains protein, vitamins, and nutrients that can lower blood pressure and reduce the risk of heart attacks.

## P. 129 EATING FOR THE PLANET

1. **b)** Choosing a plant-based source of protein, like tofu (made from soybeans), produces less greenhouse gas.

2. **b)** Picking a smaller burger means you reduce greenhouse emissions, and it's probably still plenty of food!

3. **b)** Cutting down forests to make farmland has a huge impact on the local environment, as well as reducing the number of trees helping to remove carbon dioxide from the air. It damages ecosystems, displaces animal species, and reduces the diversity of plants and animals in the area.

4. **a)** Importing food from abroad, especially by plane, has a big carbon impact. Local is better, if you can get it.

5. The answer is... it's not clear. While producing and importing livestock feed has a higher carbon footprint, and the high demand for grain can lead to deforestation in order to create land to grow it on, cattle that are grass fed still produce methane in large quantities and require huge amounts of land for grazing, which is sometimes also created by deforestation.

6. **b)** Chemical fertilizers, while they're effective, can leach into the surrounding soil and cause damage to the environment (see page 131). Farmers can collect and use animal droppings, which are rich in nitrogen and cost nothing.

7. **a)** While plant milks are all a relatively low-carbon option compared to dairy, a single almond requires 5 liters of water to grow and around 80% of the world's almonds are grown in California, where drought has become a serious problem.

8. **b)** Plastic wrappers create waste, which pollutes the environment and can cause harm to animals if they eat it or get tangled in it. While plastic can help prevent fresh food from spoiling, meaning it can be kept longer, it's easier to pick a snack which doesn't need wrapping—something that's been freshly made, or a fruit which comes with its own protective covering (peel).

9. **a)** Chicken has a significantly lower carbon footprint than lamb.

10. **b)** Plastic bottles contribute to pollution, and as long as your tap water is good to drink, it's cheaper too!

11. **b)** Food waste is a big problem; it can pollute the environment, but also the greenhouse gases that went into creating the food still have an impact. It's so easy to think ahead, and keep food waste to a minimum.

12. **a)** Concentrated juice is easier to transport. Water is evaporated out, so the concentrated juice fits in fewer trucks to be transported around. Water can be added again closer to the point of consumption.

## P. 133 CROP ROTATION

**FIELD A**
YEAR 1: Peas; YEAR 2: Cabbages;
YEAR 3: Wheat

**FIELD B**
YEAR 1: Wheat; YEAR 2: Peas;
YEAR 3: Cabbages

**FIELD C**
YEAR 1: Cabbages; YEAR 2: Wheat;
YEAR 3: Peas

## P. 136 CATTLE

**1. c)** A cow technically has one stomach, but it's divided into four sections and food passes through different parts of the stomach at different stages of digestion—partway through the process, the food is brought back up into the mouth as cud, to be rechewed by the cow's back teeth (molars) then swallowed again into a different stomach!

**2. d)** Flank is the part just before the back legs, but Frank isn't part of a cow.

**3. b)** Beef is the third most consumed meat, after chicken and pork.

**4. d)** Leaving meat to cure at a low temperature means the naturally occurring enzymes can break down the tough muscle fibers and make the meat tender. This can also be achieved by mechanical means (pounding the meat with a hammer), or by adding bromelain, an enzyme extracted from pineapples.

**5. c)** They can see all around them except a 30° wedge directly behind their head.

**6. a)** When bone pieces are heated to 700°C (1,290°F) in a sealed container without oxygen, they create a porous black substance containing no organic matter. It can be used for water purification, sugar refining, and creating ink, and has actually been used on a space satellite for heat shielding!

**7. d)** Leather is a byproduct of the meat industry, gelatin is used in many foodstuffs to make it chewy, and cow ear hairs are used in making horsehair paintbrushes.

**8. c)** About 1,700 kg (1.9 tons)—the weight of a large car.

**9. c)** Cows actually sweep grass up with their huge tongue, without using their teeth, and can scoop up as much as a five-inch square of grass in one lick.

**10. e)** Most parts of a cow end up in your food somehow! Udders have historically been eaten in parts of the UK as part of a dish called elder. Due to the risk of transmitting BSE (bovine spongiform encephalopathy, or mad cow disease), the spine and nervous system of cows is not currently considered safe to eat in many countries.

## P. 137 INSECTS AS FOOD AROUND THE WORLD

**1. d)** The witchetty grub is native to Australia. They're eaten as bush tucker—a term for foods native to the region—by Aboriginal Australians. When roasted they become crispy on the outside, like roast chicken.

**2. c)** Wasps are part of the order Hymenoptera (winged insects), and in the Yunnan province of southwest China, the larvae of several species are eaten after deep-frying. They're often sold in a wasp nest!

**3. b)** In Thailand, and across South Asia, crickets are soaked in water, cleaned, deep-fried and eaten as snacks. Thailand produces around 7,500 tons of crickets every year, and has around 20,000 cricket farms.

**4. a)** Stinkbugs are eaten in South Africa. They're collected before dawn, when they're easier to catch, and need to be caught alive so that their poisonous chemicals can be released. They're thoroughly boiled, and often served fried with a little salt.

## P. 142 NAME THAT HYBRID

Answers: **a) 6** (Zonkey), **b) 3** (Liger), **c) 1** (Wholphin), **d) 2** (Grolar/Pizzly Bear), **e) 7** (Beefalo), **f) 5** (Mule), **g) 4** (Cama)

# INDEX

# PICTURE CREDITS

**SHUTTERSTOCK**

4, 5, 16, 17: © Delpixel
4, 134, 135: © Anat Chant
11 top: © CosmoVector
11 bottom, 25 left: © Quang Ho
14, 148: © AJT
16, 17: © Irina Sokolovskaya
18, 19: © YDU Mortier
21 top: © Dibrova
21 bottom, 155: © Dudarev Mikhail
22: © Moving Moment
23, 48 top, 49 top, 121: © Nataly Studio
24, 25, 28, 130, 137 bottom left: © Irin-K
25 center: © Hong Vo
25 right: © Susii
26: © Vladimir Dudkin
27 top right: © Drakuliren
27: © KA-KA
27, 31, 98 middle: © Baibaz
28, 29: © Henrik Larsson
33: © Tanatat
34: © Nattika
35 bottom: © Pektoral
36-37: © Triff
38: © Domnitsky
39 farthest right: © Kaca Skokanova
40 left: © Alexey Smolyanyy
40 right: © Chrispo
41 top: © Thomas Dutour
41 left: © RAndrei
41 bottom right, 133 middle: © Banprik
42 top & bottom left: © Edward Westmacott
42 bottom right: © Mikeledray
43: © Morphart Creation
44 top: © Barmalini
44 bottom: © Barba Jones
45 right: © Jiang Hongyan
46: © Pernsanitfoto
48-49 bottom: © Alexander Raths
49 middle: © Gorra
52 top, 54, 55: © Amri Azhar
53 background: © Petr Baumann
53 bottom: © Robyn Mackenzie
56, 77, 89: © Railway FX
58: © Linyoklin
60 right: © Matkub2499
61: © Kaiskynet Studio
62 bottom left: © Cynoclub
62 bottom right: © Gresei
62 upper left, 111: © Vitaly Korovin
63 right: © Notsuperstar
64, 68, 150: © Maor Winetrob
64-65 (center): © Chalermchai Choychod
66, 67: © kwanchai.c
70-71: © Goldnetz
72 top, 122: © Duda Vasilii
72 left: © Sha15700
72 right: © Marilyn Barbone
73 left: © Da-ga
73 right: © Peter Vanco
74-75: © Images.etc
76: © IrinaK
77 bottom: © Prapann
79: © Tsekhmister
84: © Ivonne Wierink
86: © Paitoon
90: © Graphics RF
92, 93: © Gyvafoto
94: © Iryna Denysova
95: © Fascinadora
96: © Niradj
97 top: © Baibaz
97 bottom: © Andrii Malkov
98 left, 108: © Valzan
98 right: © Tatiana Volgutova
99 middle: © Bestv
99 right: © Cozine
100: © Afonkin_Y
101 top, 153: © Perla Berant Wilder
101 bottom: © Natdanai Srichaiyod
102, 103: © Jiffy Avril
104: © Guntsoophack Yuktahnon
105: © Winston Link
107: © Kamira
109: © Hibrida
113: © Sayam T
115 top: © Dreamloveyou
115 bottom: © Djsash
118-119: © Evgeny Karandaev
122-123: © Prostock-studio
120, 121: © BJ Photographs
126: © FloridaStock
127: © Steve Cukrov
129: © Stockcreations
130, 154: © Paphawin Laiyong
131: © Itsik Marom
132: © Unpict
133 left, 154: © Master Q
133 right: © Svitlana-ua
134, 135: © Trum Ronnarong
137 (map): © Tanarch
137 bottom middle: © Prachaya Roekdeethaweesab
137 bottom right: © Schankz
137: © Shutterstock
140: © Amri Azhar
141 bottom right: © Leungchopan
141 bottom left: © Eric Isselee
143: © Peter Etchells
145: © Itman__47

Unless otherwise stated, illustrations are by Rob Brandt. Every effort has been made to credit the copyright holders of the images used in this book. We apologize for any unintentional omissions or errors and will insert the appropriate acknowledgment to any companies or individuals in subsequent editions of the work.